3018370

Coal Age Empire

Pennsylvania Coal and
Its Utilization to 1860

By Frederick Moore Binder

Commonwealth of Pennsylvania
PENNSYLVANIA HISTORICAL
AND MUSEUM COMMISSION
Harrisburg, 1974

Contents

"Here a bountiful Providence has lavished in great profusion his richest gifts . . . Our state is an *Empire* within itself."

—Cephas Grier Childs, editor, *Philadelphia Commercial List*, from his Twenty-four Notebooks, 1845-1865, Historical Society of Pennsylvania.

Introduction

THE definitive study of a natural resource, particularly a resource as plentiful and significant as coal, would be an undertaking of enormous proportions. The writer, therefore, has chosen to examine coal principally from the standpoint of its use. He has imposed, also, further restrictions—those of time and place. This study covers the period in United States history before the Civil War and deals with the utilization of the several kinds of coal found in Pennsylvania. In pursuing the theme of the utilization of Pennsylvania coal through the varied materials of research, it seemed apparent to the writer that use was the key to the rise of the coal industry in the United States. In turn, the coal industry was a basic factor in the emergence of the new industrial nation. The study proposes to outline the development of fuel experimentation, the introduction of mechanical improvements and methods of burning coal, the marketing techniques, and the subsequent expansion of markets for the anthracite and bituminous wealth of the State.

In the period before 1860, Pennsylvania coal became a great staple of the Commonwealth, and of the nation, as experimentation and technological advancements discovered new uses for it. The United States in this period experienced the strong, young forces of the beginnings of her industrial revolution. Much of this change and development was made possible through the utilization of Pennsylvania mineral fuel.

The existence of coal deposits in Virginia, Pennsylvania, and the western country was common knowledge before the Revolutionary War. For the most part, Americans ignored the wealth of carbon energy beneath the soil and used wood to warm their homes and cook their food. At the beginning of the nineteenth century little coal was mined in the United States. Charcoal supplied the fuel needs of blast furnace and forge. Steam power, when applied at all, usually was produced by burning resinous pine knots under the boilers of the engines. The great stands of oak, hickory, and pine seemed inexhaustible. Timberland not only furnished fuel supplies for rural America, but for the towns and growing cities of the federal union. In the East the small amounts of mineral fuel which were used in the early period came from the bituminous, or soft-coal, mines of the

1

James River in Virginia, or were imported to the seaboard from Great Britain. So long as wood was abundant and cheap, little interest was displayed in native coal deposits. There were a few exceptions of course. In western Pennsylvania easily accessible outcroppings of bituminous coal, especially in the vicinity of the thriving frontier town of Pittsburgh, were used locally in home and factory. In the eastern part of the State the wild, broken country of the Wyoming, Lehigh, and Schuylkill valleys furnished anthracite, or hard coal, for the forges of some of the rural blacksmiths.

Anthracite was difficult to burn. The suspicion with which the coal was regarded outside its immediate locale was gradually replaced by confidence in its combustible properties as new grates and stoves, specifically designed to burn hard coal, were placed on the market. A new age in domestic comfort and convenience began in the towns and cities of the East. The price of wood rose with the diminishing acres of forest land. But increased tonnages of anthracite placed hard coal within the reach of all. By 1840, twenty years after the coal trade had begun, Pennsylvania anthracite no longer was regarded as a luxury but as a necessity by the urban populations of Philadelphia and New York.

Constant fuel experimentation and technological advancement before 1860 proved anthracite to be efficient in the generation of steam in stationary engines. The invention of anthracite marine boilers introduced the fuel to the steamboat and steamship. Important strides also were made in the development of the anthracite coal-burning locomotive. The iron industry provided another significant market as improvements in blast furnace and rolling mill enabled the ironmasters of the East to utilize Pennsylvania anthracite coal.

The concentration of anthracite coal deposits within the boundaries of the State made the resource a Pennsylvania monopoly. Pennsylvania produced over ninety-nine per cent of the hard coal of the United States. The rise and development of the anthracite trade resulted from the various uses of the fuel in home, industry, and transportation. Less than four hundred tons was mined and shipped to market in 1820. Forty years later the amount had risen to an annual production in excess of eight million tons.

Confined to the West by the linear ranges of the Appalachians, profitable shipments of free-burning Pennsylvania bituminous coal did not reach eastern markets until railroads penetrated the mountain barrier during the eighteen-fifties. Bituminous coal was mined from the hills overlooking Pittsburgh and from the regions of the Monongahela and Youghiogheny rivers. It became basic to the manufacturing enterprise and urban growth of the Ohio Valley. At the end of our period, in 1860, Pennsylvania mined nearly half the bituminous coal produced in the country.

By virtue of her geographic position, the Commonwealth was able to furnish mineral fuel to the populations of both East and West. Hundreds of miles of improved river navigation, canal lanes, and railroads were constructed to bring the coal to market. Between 1820 and the beginning of the Civil War, the United States began a power revolution through Pennsylvania coal.

Contemporary observers were conscious of the material changes brought about through the use of coal. Great Britain, industrial giant of the nineteenth century, constantly was cited as the prime example of what a nation could become when blessed with abundant reserves of mineral fuel. Predictions of future American economic supremacy published in the newspapers and periodicals of the day frequently were based on the development of Pennsylvania coal resources. Pennsylvanians were proud of the abundant material wealth of their State. Some maintained that through the possession of coal lands alone, the Commonwealth could hold the Union in tribute. These manifestations of Pennsylvania economic sectionalism dominate the period. But between the lines of enthusiastic and exaggerated claims, the historian is able to discern the stirrings of a new era. By 1840 America had entered what the journalists and politicians of the day chose to call the "Coal Age."

The roots of the industrial revolution in the United States are found in the period before the Civil War. The symptoms of a changing economic order are evident: the beginning of the concentration of population in cities, the increased use of steam-powered machinery, the investment of capital in industry and transportation routes, the large number of patents for technical improvements, the evolution of a skilled group of mechanics, and the material comfort of town and city dweller reflected in the domestic conveniences of central heating and gas light. Much of this would have been difficult if not impossible to attain without a source of cheap energy. Coal furnished that energy, and in turn it was Pennsylvania which furnished eighty per cent of the nation's coal.

I wish to acknowledge the courteous assistance received from staff members of the following societies: the American Philosophical Society, the Historical Society of Pennsylvania, the New-York Historical Society, the Chester County Historical Society, and the Historical Society of Schuylkill County. I also am indebted to the libraries of the University of Pennsylvania, the Sullivan Memorial Library of Temple University, the Library Company of Philadelphia, the Free Library of Philadelphia, and to the librarians of the Pennsylvania Railroad and the United Gas Improvement Company for permitting me to make use of their excellent facilities.

During the course of research, many individuals contributed suggestions which have been invaluable in bringing together the threads of the story. To each of these persons I owe a debt of gratitude. But I am particularly indebted to Mrs. Autumn L. Leonard of the Division of History of the Pennsylvania Historical and Museum Commission at Harrisburg; H. M. Irwin and W. H. Higinbotham, officials of the Delaware and Hudson Railroad Corporation in New York; R. M. Van Sant and Vernon Thomas of the Baltimore and Ohio Railroad Company, Baltimore; I. L. Gordon, Publicity Manager of the Reading Company in Philadelphia; and Fred Donaghy of Donaghy and Sons, located at 2601 Pine Street, Philadelphia, one of the oldest independent retail coal firms in existence in the city. All were kind enough to make available significant papers, reports, and minute books, without which the compilation of necessary material would have been an impossible task.

Dr. Walter L. Slifer of the Bituminous Coal Institute in Washington, D. C., Dr. Louis C. Hunter of the Industrial College of the Armed Forces, the late Dr. Howard N. Eavenson of Pittsburgh, and Professors James A. Barnes and Harry M. Tinkcom of Temple University, evidenced interest in the progress of this undertaking and offered valuable encouragement and advice.

I must also express the particular gratitude that I feel to the late Professor Arthur C. Bining of the University of Pennsylvania, who first inspired my interest in Pennsylvania economic history, guided me throughout my graduate career, and gave me the advantage of his knowledge and thorough understanding of the field. My indebtedness to him can only be surpassed by my indebtedness to my wife for her patient understanding and unfailing cooperation.

Finally, I wish to thank the members of the Pennsylvania Historical and Museum Commission, who approved the publication of my study; Mr. William J. Wewer, Executive Director, and the late Dr. S. K. Stevens, past Executive Director; Dr. Donald H. Kent, Director of the Bureau of Archives and History; and Mr. William A. Hunter, Chief of the Division of History, who supervised, and Mr. Harold L. Myers, Associate Historian, who handled publication.

CHAPTER I

Hearth and Home

NEW FUEL FOR THE EAST

THE story of the development of the Pennsylvania coal trade with its widespread domestic market is inseparable from an account of the uses of this mineral fuel. Although *The North American Review* in 1836 referred to Pennsylvania as "the Key-Stone State . . . not solely by reason of its geographical position and its magnitude but on account of its natural resources also,"[1] it was William Bigler, governor of Pennsylvania from 1852 to 1855, who used the phrase "the Key-stone of the Federal Arch" when describing the State's abundant supplies of mineral wealth. The Governor called specific attention to Pennsylvania's rich deposits of iron ore and limestone and her extensive bituminous coal seams. He then proudly noted: "Her anthracite coal beds, furnishing a choice and cheap fuel for domestic purposes, for generating steam for the stationary and locomotive engine, as well as for the propulsion of our steamships, give to her a trade almost exclusively her own."[2]

The major uses for anthracite listed by Governor Bigler were the results of many years of experimentation. Often the work was hampered or delayed by the forces of supply and demand which depended upon public suspicion or rejection, lack of capital, inadequate transportation routes, or by technological difficulties. And yet, within the span of four decades, between 1820 and 1860, Pennsylvania anthracite came increasingly to be regarded not only as a household commodity, but as a basic source of power for industry and transportation. It was through the use of Pennsylvania anthracite that much of nineteenth-century America found the energy of progress and of motion.

By water, and later by rail, coal found its way to the consumer. In spite of the early arguments concerning the varied uses of Pennsylvania anthracite in industry, the major market for this eastern fuel during the first decade of the trade was the American home. Domestic conversion to anthracite grew steadily until the coal was looked

---

[1] *The North American Review*, XLII (1836), 256.
[2] *Pennsylvania Archives*, Fourth Series, VII, 516.

upon as a household necessity. In 1834 the Pennsylvania Senate report concerning the coal trade stated, "Coal is even yet used by comparatively a small portion of our population," and observed that the use of wood fuel in New York and Philadelphia (city and county) had kept pace with population growth. The report hastened to add, however, that between 1830 and 1833 there had been a large increase in the use of coals of all kinds.[3] The dependence upon Pennsylvania anthracite fuel by a part of the population of New York was evident as early as 1831-32, when a hard-coal shortage, caused by an unusually severe winter and new demands on the supplies contracted for by dealers, resulted in considerable suffering in that city.[4]

The popular utilization of anthracite in the home and in public buildings was brought about by a deluge of informative propaganda by the friends of anthracite coal; new and practical inventions in grates, furnaces, and stoves; and through scientific fuel analysis. Experiments in anthracite heating and later in cooking were encouraged by men of enterprise. Motivated by desires for profit and pride in accomplishment, capitalist, speculator, retailer, and politician enlisted the aid of scientist, geologist, political economist, and journalist in a successful campaign to educate the public in the virtues of anthracite, "the most despised of the combustibles."

Virginia bituminous coal from the James River mines had been available to the West Indies and the eastern seaboard since 1758. Ten years later Great Britain began exporting small amounts of soft coal to the continental colonies and the islands.[5] Bituminous coal burned easily and never presented the combustion problems of anthracite, but as long as wood remained plentiful and cheap, soft coals were not consumed in any quantity by the eastern towns, with the exception of Richmond. Wood continued to be the chief household fuel of the eastern population centers until the rise of the anthracite trade in the decade of the eighteen-twenties. Dr. James Mease, describing Philadelphia in 1811, wrote:

The principal article of house fuel in Philadelphia, is hickory, oak or maple wood. Pine wood is used chiefly by brick

[3] *Report of the Committee of the Senate of Pennsylvania upon the Subject of the Coal Trade* (Harrisburg, 1834), p. 43.

[4] *Miners' Journal,* December 17, 1831; March 31, 1832.

[5] H. N. Eavenson, *The First Century and a Quarter of American Coal Industry* (Pittsburgh, 1942), pp. 32-36; C. E. Peterman, "Early House Warming by Coal Fires," *Journal of the Society of Architectural Historians,* IX, December (1950), 21-24.

burners and bakers. Coal is only partially used in dwelling houses, but would be in general demand for counting rooms, offices and chambers, owing to the danger from fire being thereby lessened, if it could be afforded at a rate as cheap as wood. The time is anxiously looked forward to, when the inexhaustible bodies of excellent coal with which our western counties, and Wayne county abound, will be brought down to Philadelphia, by means of that great chain of inland navigation, which has been so long in contemplation, and by the removal of the obstructions in the Lehigh.[6]

During the War of 1812, Dr. Mease, then secretary of the Philadelphia Society for Promoting Agriculture, received specimens of anthracite from Luzerne County sent by Jacob Cist, of Wilkes-Barre, an early entrepreneur of the anthracite trade. Cist urged a canal between the Delaware and the Susquehanna so that Philadelphia could be supplied with gypsum and anthracite. The latter, shipped by way of the Susquehanna to tidewater and thence coastwise to New York City, had found there a limited but promising market. Cist, undoubtedly interested in promoting a Philadelphia household market for his anthracite, touched upon a theme that was to prevail in the arguments of the coalmasters for many years. Fuel, cheap and excellent, was no less important to men than water, argued Cist. Here was opportunity to aid the needy (and to make a profit at the same time).

How many miserable wretches, who shiver over your wood fires, which cost them 6 and 8 dollars per cord, could be made comfortable at half the price? Much of the coal from here is shipped at Havre de Grace or at tide, for New York, and readily commands 50 per cent per bushel more than the best Liverpool coal.[7]

How well the pen of the wily Franklin could have served the anthracite interests, for if Franklin was not the first practical scientist and propagandist in home heating, he certainly was the most famous. Desirous of little more than a warm home to keep him and his family safe from the chill of winter, Benjamin Franklin conceived the "Pennsylvanian Fireplace" or Franklin Stove. He did not patent the invention, but gave the model to Robert Grace, who then manufac-

---

[6] James Mease, *The Picture of Philadelphia* (Philadelphia, 1811), p. 125.
[7] *Memoirs of the Philadelphia Society for Promoting Agriculture* (Philadelphia, 1814), III, 141.

tured these cast-iron fireplaces at his Chester County furnace.[8] In order to help Grace sell the product, Franklin turned pamphleteer, and through the subtlety of a learned treatise the sage peddled his contrivance of comfort. Cold drafts or great, bright fires were dangerous to the health of all, said Franklin. The ladies, in particular, were constant victims of "rheums," defluxions resulting in loss of teeth and shriveled skin caused by improper heating apparatus. After the artful appeal to woman's vanity, he listed fourteen advantages of his invention and gave detailed instructions how to install it. He then closed with the most conclusive argument of all: the rooms were warmer with one-quarter the amount of fuel.[9] The "Pennsylvanian Fireplace," invented in 1740, was designed to burn wood, the common fuel of the colonies. Later, Franklin, while in England, perfected a stove which burned soft coal and consumed much of the annoying smoke.[10] When Pennsylvania hard coals were tried in Franklin's inventions and in the old ten-plate stoves during the early years of the anthracite trade, incomplete combustion resulted in waste and expense.[11]

Prior to 1800, anthracite was used in many smiths' forges in the valleys of eastern Pennsylvania. In the year 1788 Jesse Fell, of Wilkes-Barre, had experimented with anthracite in his nailery and found it to be a cheap, profitable fuel. The inhabitants of Wilkes-Barre firmly believed that it could not be used as a domestic fuel, since without a forced draft, its fire would go out. This local opinion had been sustained by a trained scientist of the University of Pennsylvania, James Woodhouse. In 1805 Dr. Woodhouse had tested Lehigh anthracite. His experiments, the first of their kind in the United States, led him to the conclusion that anthracite was superior to Virginia bituminous when long, continued periods of heat were needed. Anthracite could be used to advantage in distilling, in evaporating large quantities of water, in melting metals, subliming salts, generating steam, and also for washing, cooking, and home heating, "provided the fireplaces are

---

[8] Carl Van Doren, *Benjamin Franklin* (New York, 1938), p. 117. For an excellent, well-illustrated popular history of early heating apparatus see J. H. Peirce, *Fire on the Hearth, the Evolution and Romance of the Heating Stove* (Springfield, 1951). The oldest known model of the Franklin stove is now in the possession of the Bucks County Historical Society.

[9] Van Doren, *Benjamin Franklin,* pp. 141-142.

[10] *Ibid.,* p. 728.

[11] J. T. Scharf and Thompson Westcott, *History of Philadelphia* (Philadelphia, 1884), III, 2271.

constructed in such a manner as to keep up a strong draught of air."[12]

Jesse Fell was not one to bow to local opinion. What is more, he probably remained ignorant of Woodhouse's experiments. In 1808 Fell constructed a crude, ten-inch iron-rod grate, stumbled upon the principle of the minimum draft, and began a revolution in home heating. Many years later he described his experience with anthracite in the following manner:

> I had for some time entertained the Idea that if a sufficient body of it was ignited it would burn; accordingly in the middle of the month of February 1808 I procured a Grate made of Small Iron rods, 10 inches in depth and 10 inches in height, and I set it up in my common room fireplace and on first lighting it found it to burn excellently well. This was the first successful attempt to burn our Stone Coal in a Grate, so far as my knowledge extends.[13]

There is evidence that Fell was not the first to contrive the use of the grate, for as early as 1800 the ingenious Oliver Evans succeeded in burning anthracite in an open grate without an artificial draft.[14] Dr. Thomas C. James heated his living room with an anthracite fire during the winter of 1804 and continued to use hard coal for more than twenty years. He predicted that someday anthracite would become the domestic fuel of Philadelphia.[15]

It was Fell's grate, however, which gained popularity in the Wyoming Valley. And it was the principle of this grate which contributed to the first successful venture into the anthracite coal trade in Pennsylvania. Abijah Smith and Company, of Plymouth, realized that in order to sell the coal the method of burning it must be sold first. The brothers Abijah and John, who formed the company, appeared in Columbia in the spring of 1808. They brought with them several tons of coal and some skilled masons who constructed Fell's grates in public places. The Smiths then gave demonstrations on the burning of Wyoming anthracite. Convincing the skeptics with these public exhibitions, the pioneer company found a small market for its fuel in the towns along the Susquehanna. Some of the coal moved on

[12] James Woodhouse, M.D., Experiments and Observations on the Lehigh Coal," *The Philadelphia Medical Museum*, I (1805), 444.

[13] Historical Society of Pennsylvania, Jesse Fell to Johnathan Fell, Wilkes-Barre, December 1, 1826. Society referred to hereafter as HSP.

[14] *Manufactures of the United States in 1860; Compiled from the Original Returns of the Eighth Census* (Washington, 1865), p. cixx.

[15] HSP, T. C. James, M.D., A Reminiscence, MS.

to Baltimore and by sea to New York. Although others joined the enterprise, by 1820 the coal trade of the Wyoming region had totaled only ten thousand tons.[16]

Public demonstrations and sworn testimonials were common devices of the early coal traders. When in 1814 Lehigh coal was hawked in the streets of Philadelphia for fourteen dollars a ton, its virtues were advertised on handbills in German and English and by affidavits signed by Philadelphia blacksmiths who had been prevailed upon, with difficulty, to use the coal. Jacob Cist, who had shown such concern for the "miserable wretches shivering over . . . wood fires" in Philadelphia, hauled a model stove from door to door, begged home owners to permit him to test anthracite in stoves constructed to burn Liverpool coal, and, for good measure, bribed journeymen in blacksmith shops to use his Lehigh coal.[17]

News of the Peace of Ghent ended the War of 1812. With the peace, Virginia and English bituminous returned to the market and ended the early, almost unnoticeable anthracite trade of Philadelphia and environs. Unless anthracite could be made available through reasonable transportation rates, it could not hope to compete with wood or the cheaper, free-burning, water-borne bituminous. Inadequate and expensive transportation to potential markets, coupled with prevailing ignorance and suspicion of this useful fuel, created the twin obstacles of price and prejudice.

When the improvements in navigation on the Lehigh and Schuylkill began bringing anthracite to tidewater, the eastern market expanded and prices adjusted gradually. Anthracite producers, individual and corporate, made a fetish out of the quest for new uses for their product, particularly in the many branches of manufacturing. But during the eighteen-twenties and early eighteen-thirties, they remained conscious of the fact that the demand for anthracite was dependent chiefly upon household consumption. Producers eagerly encouraged technical improvements in grates and stoves and welcomed chemist and inventor to their ranks.

While the coal trade was still in its infancy, a champion of anthracite coal, Marcus Bull, conducted "experiments to determine the

---

[16] William Griffith, "The Proof that Pennsylvania Anthracite Coal was first Shipped from the Wyoming Valley," *Proceedings and Collections of the Wyoming Historical and Geological Society*, XIII (1913-14), 65-70.

[17] Charles Miner to Samuel Packer, November 17, 1833, from the *Report of the Committee of the Senate of Pennsylvania upon the Subject of the Coal Trade*, p. 95; J. L. Bishop, *A History of American Manufactures from 1608-1860* (Philadelphia, 1864), II, 185.

comparative quantities of Heat, evolved in the combustion of the principal varieties of Wood and Coal, used in the United States, for fuel." His paper was read before the American Philosophical Society in April, 1826, and was summarized in volume one of the *Journal of the Franklin Institute,* then known as *The Franklin Journal and American Mechanics Magazine.*[18] The article began with a note of pessimism for all coal users and dealers: "The principal article of fuel used in the United States is forest wood, which, from necessity, or choice, will continue to be so, in many sections of the country, notwithstanding the abundant supply of anthracite and bituminous coal, already discovered in some of the states." Anthracite was difficult to burn in open grates. Until this problem was overcome, said Bull, there could be no general introduction of anthracite for home heating. This early heating engineer made some encouraging suggestions on the improvement of grates then in use. Contrary to popular belief, anthracite did not need a strong draft to burn effectively. A rush of air would destroy the fire as it tended to reduce the temperature and hamper combustion. Deeper grates and enclosed ash pits to heat the light draft would solve the basic problem. He concluded his lengthy treatise with the plea that furnace manufacturers read his statements with care.

Marcus Bull was convinced that anthracite possessed great utility in the arts as well as in the home if burned under proper conditions. When confronted with experiments favoring bituminous and wood fuels, he defended himself in an interesting manner.[19] Bull repeated his earlier findings and supported his contentions with public testimony in a pamphlet issued by the Lehigh Coal and Navigation Company.[20] Anthracite, argued the scientist, had been proven not only in the laboratory, but in shop, forge, and home, to the delight of users. His remarks remain an unusual combination of scientific investigation and blatant pamphleteering. Consciously or unconsciously serving the interests of anthracite coal, Marcus Bull's experiments and writings contributed to the strengthening of the early anthracite market.

---

[18] Marcus Bull, "Experiments to determine the comparative quantities of Heat evolved . . . ," *Transactions of the American Philosophical Society,* III, New Series (1830), 1-63; *Journal of the Franklin Institute,* I (January-June, 1826), 285-289.

[19] Marcus Bull, *An Answer to "A Short Reply to 'A Defence of the Experiments to Determine the Comparative Values of the Principal Varieties of Fuel, etc.'"* (Philadelphia, 1828).

[20] *Facts Illustrative of The Character of The Anthracite or Lehigh Coal* (Philadelphia, 1824).

Ideas for anthracite stoves, ovens, and furnaces, singly or in various combinations, developed side by side with improved grates for open fires. The greatest boom was between 1828 and 1835, but applications for patents continued year after year to the Civil War.[21] A prolific inventor in this realm of comfort economy was the remarkable Reverend Doctor Eliphalet Nott. A century ago the name of Eliphalet Nott was quite familiar, for he was one of the few significant figures, outside the realm of politics, who frequently entered the public eye.[22] President of Union College for sixty-two years, he attained a national reputation as an educator, pulpit orator, ardent prohibitionist, and practical inventor. Nott's first patent for a rotary grate for burning anthracite and for shaking the ashes was granted in March, 1826, revised and improved in 1828 and again in 1832.[23] He turned his attention to anthracite stoves and in June, 1833, applied for and was issued eleven patents on an improved coal stove.[24] Over the years Nott patented thirty different types of stoves and was the chief power behind H. Nott and Company, leading stove manufacturers of the day.[25]

Grate and stove manufacturers proliferated in the eastern cities. The more rigorous the climate, the more numerous and prosperous they seemed to become. One would offer a grate of Boston origin, guaranteed to eliminate dust and falling ash;[26] another would announce the newest virtues of his particular coal stove. Advertisements of this kind dotted the pages of the newspapers in the major coastal cities of the Northeast.[27] Hazard's *Register of Pennsylvania* encouraged the use of anthracite stoves in 1828,[28] the same year the first anthracite stove manufacturers, Williamson and Paynter, opened

---

[21] See *Journal of the Franklin Institute,* IX (January-June, 1830), 136, 145-148; X (July-December, 1830), 78; XV (January-June, 1833), 99-100, 172-174; XVI (July-December, 1833), 87, 106-108, 404-407; XVII (January-June, 1834), 395; XVIII (July-December, 1834), 200-202; XIX (July-December, 1835), 205; and volumes XX (July-December, 1835) to LXX (July-December, 1860); *7th Annual Report of the American Institute of the City of New York* (1849), p. 119.

[22] D. R. Fox, *Dr. Eliphalet Nott (1773-1866) and the American Spirit, a Newcomen Address* (Princeton, 1944).

[23] *Journal of the Franklin Institute,* XV, 172-174.

[24] *Ibid.,* XVI, 404-407.

[25] Fox, *Eliphalet Nott,* p. 11.

[26] Samuel Hazard (ed.), *The Register of Pennsylvania,* XIV (July-December, 1835), 141.

[27] See various issues, 1830 to 1860, of the following newspapers: *The Pennsylvanian, Philadelphia North American, Public Ledger, Boston Daily Advertiser, New York Commercial Advertiser.*

[28] Hazard, *Register of Pennsylvania,* I (January-June, 1828), 312.

their doors in Philadelphia.[29] "By adopting stove furnaces and pipes, they can dispense with chimnies and fireplaces, and the removal of soot and obstructions by sweepings will not be required."[30] To Samuel Hazard and to many New Yorkers, Philadelphians, and Bostonians who could afford the new fuel and the new apparatus, the millennium in home comfort had arrived.

Although heating stoves never completely took the place of the open grate in the period before the Civil War, they became common equipment for home heating in the urban areas. Central heating was slower to develop, but the concept was an old one. Daniel Pettibone had invented a "rarefying air stove" or hot-air furnace in 1810. The furnace was adopted by some home owners and installed in a few public buildings. The Philadelphia Bank, the Almshouse, St. Augustine Roman Catholic Church in Philadelphia, and the House of Representatives in Washington used Pettibone's apparatus. This was a wood-burner, and it remained for Professor Walter R. Johnson of the Franklin Institute to apply anthracite to the air furnace in 1825. Johnson, later to become one of the nation's authorities on fuel analysis, placed his furnace in the cellar and piped out the smoke and gases through drums which penetrated the first, second, and third stories.[31] Johnson's furnace was adopted in public buildings when the smaller coal stoves proved troublesome and expensive. The Eastern State Penitentiary at Philadelphia installed this type of furnace and warmed twenty cells simultaneously.[32] One may assume that by the decade of the fifties, most public buildings utilizing mineral coal for fuel used an adaptation of the air furnace in central heating. Many did, including the House of Refuge at Philadelphia and the Treasury buildings in Washington.[33] But the humble grate remained popular. As late as 1859 the proprietors of New York's Fifth Avenue Hotel, the "Palace of the People," announced with pride that each room had been equipped with a modern open fireplace grate perfected by a Pittsburgh manufacturer, and called attention to the aesthetic appeal of an open coal fire filling the room with warmth, "mirth and sociability."[34]

---

[29] Scharf and Westcott, *History of Philadelphia*, III, 2272.

[30] Hazard, *Register of Pennsylvania*, I, 312.

[31] Scharf and Westcott, *History of Philadelphia*, III, 2271.

[32] *Pennsylvania House Journal*, 1829-30, II, 549.

[33] *Pennsylvania Legislative Documents*, 1857, p. 718; *Senate Executive Document*, No. 31, 32nd Congress, 2nd Session, contains an item regarding the fuel used in the Treasury buildings in Washington during 1852: 149 tons of anthracite, 158 tons of bituminous.

[34] *Scientific American*, I, New Series, July 2, 1859, p. 6.

It was not long after the acceptance of Pennsylvania anthracite as a clean, efficient fuel for "store and parlour" that kitchen fuel experiments began. Two Philadelphians, James Vaux and Thomas Mitchell, among the first to proclaim the advantages of anthracite coal grates in home heating, recommended open kitchen grates for cooking and offered original designs to the public. Mitchell used a slip grate to regulate the amount of coal. A top of ordinary sheet iron radiated the heat of the fire. "No stoop, no smoke, no odors: little care and less fuel," might have been the slogan of these early practical improvers, for their articles noted the above advantages, in more dignified manner, of course.[35] The editor of the *Journal of the Franklin Institute,* commenting upon Mitchell's grate, forecast a "cook's revolution" in London should British culinary artists be required to use wood after burning bituminous coal. The advantages of anthracite over bituminous were well known. Concluded the editor, "We believe that the time will soon arrive, when our servants will, if required to use wood for cooking fires, object to it on account of the difficulty in managing it."[36]

Elaborate grates, often necessitating the revamping of chimneys, were set permanently in kitchen fireplaces of some homes and hotels in Philadelphia, New York, and Boston. Anthracite was the fuel of the fashionable, who turned not a coal themselves, but hired servants to prepare fire and food.[37]

The mine owners were confident of a great increase in anthracite consumption by the introduction of their fuel for culinary purposes. The *Miners' Journal* voiced the opinion that in 1829 not one in ten knew anything about anthracite for cooking; but by 1830 there wasn't one in ten who had not become an authority on the subject.[38] Consumption did increase, but Schuylkill County producers were not satisfied with the kitchen grate. "The expensive and complicated contrivances hitherto presented to the public for cooking with anthracite coal, have been a great barrier to the introduction of this economical fuel into the culinary department," said their trade organ. The large cities were urged to take note of Pottsville inhabitants eating their steaming dinners, hot off the coal fires burning under two

---

[35] *Journal of the Franklin Institute,* II (July-December, 1826), 293-295; III (January-June, 1827), 121-122.

[36] *Ibid.,* III, 123-124.

[37] *New York Evening Post; Boston Courier.* Reprinted in the *Miners' Journal,* April 28, 1827 and October 18, 1831.

[38] *Miners' Journal,* March 13, 1830.

hundred anthracite grates. These grates were simple and economical, the *Journal* stated, selling for about ten dollars each.[39] Mine operators in Pottsville subscribed $100 and donated it to the Franklin Institute to be used in encouraging the invention of an anthracite cook stove, price not to exceed ten dollars. The Franklin Institute offered the cash inducement and promised a silver medal to the winner.[40] The suggested price of ten dollars was difficult to meet, and Hazard's *Register* continued to lament the high cost of all anthracite appliances. Anthracite coal stoves, kitchen grates, and simple baking ovens should be put on the market within the reach of all.[41] The Board of Managers of the Fuel Savings Society of Philadelphia announced that anthracite should be the fuel of the poor. This would be possible only if a cheap apparatus were constructed for burning coal in common room and kitchen. The board set the price of an anthracite stove at six dollars and promised a dividend to anyone inventing such an economical unit.[42] New York formed a fuel committee which posted a prize of fifty dollars for a cheap cooking grate utilizing anthracite coal. Schofield and Hall, of Poughkeepsie, won the award and that season promised to market six hundred to one thousand anthracite kitchen grates retailing for about four dollars each.[43]

Stimulation of new and cheaper technical improvements increased demand. At the same time, better transportation and competition with other fuels, especially wood, brought down the price of coal when the mineral fuel was in good supply. The cost of coal, like the cost of wood, over the years, was influenced by a multitude of variables, among them supply, demand, kind, area, and climate.[44] New York consumed over 50,000 short tons of anthracite in 1832-33 worth $513,797, to almost 266,000 cords of wood valued at $615,914.[45] In January, 1836, anthracite averaged between eight and ten dollars in New York City, while hickory wood, delivered, brought $3.25 per cart load and oak $2.50.[46]

---

[39] *Ibid.*, August 6, 1831.
[40] *Ibid.*, March 19, 1831; *Journal of the Franklin Institute*, XII (July-December, 1831), 1-2.
[41] Hazard, *Register of Pennsylvania*, VII (January-June, 1831), 238; for a similar article see also *Poulson's American Daily Advertiser*, October 15, 1831.
[42] Hazard, *Register of Pennsylvania*, VII (July-December, 1831), 127-128.
[43] *New York Evening Journal*. Reprinted in the *Miners' Journal*, September 17, 1831.
[44] See Table A, Appendix, for average anthracite prices in the New York market.
[45] Hazard, *Register of Pennsylvania*, XI (January-June, 1833), 289.
[46] *New York Commercial Advertiser*, January 13, 1836.

Prices in Philadelphia varied as much as in New York.[47] *Poulson's Advertiser* in 1831 estimated fuel consumption for an inexpensive anthracite stove for one season to be about $4.50 and the cost for firewood to heat the open hearth to be twenty-one dollars a season in Philadelphia. Small wonder the coalmasters pressed for cheap grates and stoves. During the cold fall of 1829, anthracite on the line of the Schuylkill was scarce because of better prices at tidewater. Most of it was shipped to Philadelphia, where it brought nine dollars a ton. The *Reading Chronicle* growled over being ignored by the passing coal boats, but took comfort in hickory wood selling for four dollars a cord. "Huzzah for Old Hickory," cheered the disgruntled *Chronicle*.[48]

One of the early retail coal yards in Philadelphia, J. Donaghy, sold anthracite egg for $3.00 per ton in 1837, $4.00 in 1845, and $4.50 per ton by 1853.[49] The Boltons at Callowhill Street wharf retailed Schuylkill coal from 1827 to 1830 for six dollars per ton, excluding cartage.[50] In the fifties, J. F. and S. Jones, of Germantown, held their prices between four and five dollars on egg, nut, and stove, but with inflation during the last years of the Civil War, Philadelphia retail coal prices soared to as high as fourteen dollars per ton.[51]

Despite price inconsistencies, the coal trade of the State expanded rapidly. Anthracite production rose from less than 80,000 tons in 1827 to approximately 258,000 tons in 1831.[52] This amount fell far short of the demand during the winter of 1831-32 in the two great anthracite marts of Philadelphia and New York. The coal operators had completely misjudged the market. In order to meet the new demands production reached the half-million mark the following year.[53] What had happened in 1831-32 to produce a shortage and to stimulate the mine owners to all but double their output for the following season? Schuylkill coal, a drug in the Philadelphia market at five dollars during the mild fall days of 1831, couldn't be had at double its usual price by late November. The consumption by the little towns of Connecticut alone was equal to that of Philadelphia four years be-

---

[47] Anne Bezanson, R. D. Gray, and Miriam Hussey, *Wholesale Prices in Philadelphia, 1784-1861* (Philadelphia, 1936), parts I & II, *passim*.

[48] Reprinted in the *Miners' Journal*, October 31, 1829.

[49] Donaghy and Sons, Philadelphia, coal yard receipts.

[50] Historical Society of Schuylkill County, Eyre-Ashurst Papers. Society is referred to hereafter as HSSC.

[51] F. and S. Jones, Germantown, Ledgers.

[52] Eavenson, *American Coal Industry*, p. 498.

[53] *Ibid.*

fore.[54] The demand on the mines was unforeseen. Reserve supplies soon were exhausted. When winter closed the water routes, the Northeast experienced its first anthracite coal shortage.

Ignoring contributory causes for the shortage, Philadelphia fixed the blame on rapid adoption of anthracite for cooking.[55] Purchases had been light that fall and independent operators gauged production according to contracts with coal agents. The large corporations, too, were hesitant to overproduce. With the chill of early winter and continued bitter weather, anthracite not only was needed to feed the new kitchen grates, but to supply the parlor stove. Another reason for the shortage was that manufacturers had begun to use anthracite for steam in larger and larger amounts. The combination of severe weather plus the concentrated efforts of the anthracite interests to increase the popularity and the uses of their fuel in home and factory descended upon the coal market with a suddenness that was unexpected.

Many New Yorkers resented the coal shortage and accused the mine operators of speculation. Congress was petitioned to repeal the duty on foreign coal. Some "traitorous" Philadelphians supported the New York petition.[56] There was legitimate foundation for complaint. Although Lehigh prices remained fairly steady, Schuylkill operators and the Delaware and Hudson Company of the Lackawanna district sold low when the market was slow and boosted prices when the shortage ensued. Prices soared to ridiculous heights. Anthracite brought as much as sixteen dollars per ton of two thousand pounds on the New York market, the highest price which was to be recorded in thirty-five years of the New York anthracite coal trade.[57] Part of the blame for the price rise should be placed on the boatmen who, like the mine owners, took advantage of the rising market and increased their freight charges.[58] By spring the coal shortage had eased in Philadelphia and anthracite sold on the average of seven dollars per ton, but the Pennsylvania fuel was still scarce in New York and that which was available brought twelve dollars a ton to the coal dealers in the city.[59]

---

[54] *Philadelphia Saturday Bulletin*, November 26, 1831. Reprinted in the *Miners' Journal*, December 3, 1831.

[55] *Ibid.*

[56] *Miners' Journal*, January 7, 14, 28, 1832.

[57] *House Executive Document*, 38th Congress, 1st Session, VI (1863-1864), 362-401.

[58] *Miners' Journal*, March 17, 1832.

[59] *Bicknell's Reporter*, April 9, 1832. Reprinted in the *Miners' Journal*, April 28, 1832.

In scarcity or in plenty the new fuel was accepted on an ever increasing scale. Domestic consumers became aware of the various kinds of anthracite on the market and made purchases with care. Dealers seldom advertised "anthracite coal," but used the names of the district or mines from which the coal came. Lehigh, Schuylkill, or Lackawanna coal brushed against Liverpool and Virginia coal in the urban markets. More specifically, red-ash and white-ash coals were noted and prices varied with the brands.[60] An inferior quality of surface coal brought to the New York market in 1829 by the Delaware and Hudson Canal Company was quickly detected by the public, whose confidence in this new anthracite was shaken immediately. It took the company an entire season to counteract this unsavory beginning.[61] *The New York Constellation* warned the coal buyers against unscrupulous dealers who would cheat them by short measure or adulteration with slate. In very poor poetry, the *Constellation* offered words of advice to its readers. Some of the lines of this bit of practical humor are worth quoting:

Rules for Buying and Burning Anthracite Coal
Done in Verse

But if your coals a quick ignition take!
And being lighted, show a lambent flame,
Of yellow, orange, or rose colored taint,
Still playing calm and gentle o'er the surface,
Like smiles upon the gentle face serene;
And if the ashes prove, instead of white,
A reddish brown, soft, fine, impalpable;
And if the fire, once lit, continue long,
Glowing and lively, sending forth the heat;
The coal is good and fit to warm the hearths,
Of honest men. Make haste to purchase more,
If more there be, and you are not supplied.

The poem, if it can be called that, rambled on. Use of grate and poker was explained in detail. Cautious application of the poker was advised. An anthracite fire burned best when let alone.

Beware of pokers—they of flesh and blood—
Nor on their iron namesakes let them lay a hand . . .

---

[60] For examples see *Poulson's American Daily Advertiser,* January through February, 1831.

[61] *Annual Report of the Board of Managers of the Delaware and Hudson Canal Company to the Stockholders* (1831), p. 5. Referred to hereafter as *Delaware and Hudson Company Annual Report.*

Even the size of the coal was put to verse: the proper measurement of coal lumps always should be equal to the egg of "that wise but slandered fowl whose timely noise saved Rome."[62]

New Yorkers preferred the red-ash anthracite from the Peach Mountain and Peach Orchard mines of Schuylkill County and later from the rich Swatara regions of Pine Grove. Its price was slightly more than the common white-ash because of increased mining costs due to the fact that the coal was found in thin seams and was difficult to work. Home owners using open grates found that the ashes did not rise rapidly when the fire was shaken. Easy to ignite, red-ash burned best in open grates. Some contended it did not chap the hands or take the varnish from fine furniture as did the white-ash. Red-ash for the open grate remained the favorite coal of Father Knickerbocker for more than thirty years.[63] White- and grey-ash anthracite were popular fuels for the closed stoves and furnaces. Lackawanna coal first catered to the domestic market, but found its largest sale in manufacturing and later as steamboat fuel.[64]

A hasty glance at medical advertisements in the city newspapers or town journals published in the middle period provides proof enough that Americans a little more than a century ago, like Americans today, were vitally concerned with their physical well-being. It was not difficult to find many citizens deeply troubled over the effects of anthracite fires upon health. From time to time accounts were printed on the near-fatal effects of escaped coal gas caused by improperly installed stoves. In 1831 the Common Council of the city of New York narrowly missed annihilation when fires were started with the valves closed in the three anthracite stoves used to heat the council chambers.[65] Henry C. Carey, formerly an outstanding free-trader who became the paragon of protectionism and patron saint of the high tariff, was gently chided by a friend for paying so little heed to his health:

> Perhaps your not being quite as well as usual may not be wholly owing to writing. I dare say it [is] in part from this occupation, being cooped up in a room with a temperature, from anthracite coal (dry and unwholesome) somewhere near 75 or 78. You ought to hear our cousin Charlotte talk

[62] Reprinted in the *Miners' Journal*, February 25, 1832.

[63] *New York Journal of Commerce*, September, 1835 to January, 1836; *American Railroad Journal*, VII, August 15, 1841; *Miners' Journal*, September 26, 1846; Colorado State Historical Society, William Jackson Palmer Papers, Coal Notes, August 21, 1856. Society is referred to hereafter as CSHS.

[64] W. R. Johnson, *The Coal Trade of British America* (Philadelphia, 1850), p. 178.

[65] *Poulson's American Daily Advertiser*, February 3, 1831.

against the hard coal fires, and tell of the injury thereby done to Eye-sight, [hair,] complexion and nerves.[66]

Whether or not Carey heeded these warnings after returning from an ocean voyage is difficult to say. Since Carey gained considerable revenue from coal lands, one doubts that he quenched his coal fires, but concludes instead that he continued to warm his bones and pen his tracts before the cheerful glow of an anthracite stove.

But the opinions of Miss Tyler and Cousin Charlotte, if not common, were at least shared by more than a few. One irate New Yorker unleashed a tirade against the anthracite furnace, crying out that this curse of mankind should be abolished:

> . . . furnaces in private dwellings will hourly destroy the health of our women and children. Hot air, with delicately or indelicately bare shoulders, bare arms, and thin clothing, contributes largely to our cemeteries, increases the number of bald heads, decayed teeth and black-craped hats . . .[67]

Experiments were conducted in the thirties to determine the extent of the effects of anthracite on health.[68] Inquiry had not ceased even after the Civil War. Dr. George Derby, surgeon to the Boston City Hospital and Harvard professor of hygiene, wrote a small book on home heating and the effects of anthracite coal fires on health. He concluded that it was not the lack of moisture in anthracite-heated homes which had a depressing effect, but the escape of noxious gases, namely colorless and odorless carbonic oxide gas (carbon monoxide). Headaches, listlessness, and dullness were the result. Death could be induced by prolonged exposure to the fumes. Derby prescribed a remedy: install a wrought-iron furnace as gases seeped from the pores of cast-iron stoves when overheated.[69]

For every critic extolling the salubrious advantages and restful beauty of a crackling wood fire and condemning the hazards of anthracite, there were a thousand champions of the mineral fuel. By the mid-forties wood had lost the battle and there was little doubt that anthracite was looked upon as a necessity for the homes of rich and poor in the cities and towns of the Northeast.[70] An article in

---

[66] HSP, Edward Carey Gardiner Collection, Henry C. Carey Papers, Caroline Tyler to Henry C. Carey, May 4, 1857.

[67] J. C. Battersly to editors, *Scientific American*, I, July 9, 1859, p. 20.

[68] H. D. Rogers and A. D. Bache, "Analysis of some of the Coals of Pennsylvania," *Journal of the Academy of Natural Sciences*, VII (1834), 3-12.

[69] George Derby, M.D., *Anthracite and Health* (Boston, 1868), pp. 1-76.

[70] *The Merchants' Magazine and Commercial Review*, XXXII (January-June, 1855), 256; *Annual Report of the Chamber of Commerce of the State of New York for the Year 1859* (New York, 1860), p. 64.

Freeman Hunt's famous *Merchants' Magazine* eulogized the anthracite of Pennsylvania and urban dependence upon it: "Commerce is President of the Nation and Coal her Secretary of State." Emphasizing that bituminous coal was little used and almost unknown among domestic buyers of the northern seaboard cities, the article continued in that wonderful, melodramatic style employed by the journals of the day:

> We could do without the gold of California, for it does not add a single real comfort to the life of man; but we could not do without our coals. The Kooh i-noor diamond is valued at $2,500,000—a sum which could purchase 500,000 tons of coal. If this diamond was dropped into the depths of the sea and lost forever, no one in the world would suffer for a single useful article the less; but if 500,000 tons of coal were prevented from coming to New York city this summer, 200,000 people would be reduced to a state of intense suffering during the next winter. Coals, then, are the real diamonds of our country.[71]

The Western Firesides

Those who dwelt along the banks of the Monongahela River, which flows through the broken hills of the western country from Virginia north to Pittsburgh, had warmed themselves in front of bituminous coal fires before the beginning of the nineteenth century. Travelers, pushing their way over the Alleghenies to the Forks of the Ohio, if they kept any records at all, usually remarked on the abundance of surface coals.[72] Kicked up by wagon tracks, uncovered in the digging of cellars and wells, or etched in black outline by persistent erosion of a mountain stream, the readily combustible fuel was dug from the earth with comparative ease. By 1800 an English traveler, John Bernard, referred to Pittsburgh as a "smokey city," recalling to him "many choking recollections of London."[73] Smoke from numerous domestic firesides bore grimy witness to coal's early triumph over Pittsburgh's winter climate.

Here, in a country which was one continuous forest in the early eighteen-hundreds, wood could be procured for practically nothing. There was not a landowner at Pittsburgh who was not ready to sell a cord of wood for half the price of coal if the purchaser would be

---

[71] *Merchants' Magazine*, XXX (January-June, 1854), 248-249.
[72] Eavenson, *American Coal Industry*, pp. 155-170.
[73] H. N. Eavenson, "The Early History of the Pittsburgh Coal Bed," *The Western Pennsylvania Historical Magazine*, XXII (1939), 167-174.

willing to cut, split, and haul it home himself.[74] Population was
scarce, labor dear, and time precious in the busy western towns. Coal
mined from Coal Hill, overlooking Pittsburgh, could be delivered
by wagon to the door for as little as five cents a bushel. This was
cheap enough, and far more convenient than cutting your own wood.
By 1808 most of Pittsburgh's six hundred soot-stained dwellings
burned coal. For two dollars or forty bushels of coal, two fires, one
for cooking and one for warmth, could be kept going about a month.
Even the poorest had two fires.[75]

The tiny towns and villages springing up in western Pennsylvania
burned bituminous as home fuel when it was easily accessible. In the
more sparsely settled sections, the knowledge of local coal deposits
often stirred little enthusiasm for home use. Westmoreland County,
which was to count coal as its source of wealth after the Civil War,
mined little up to 1850, and wood remained the predominate domestic fuel to 1860.[76] In some parts of the western coal country and the
Ohio Valley, prejudice against coal for cooking purposes continued for
many years. There was little question of special grates, or even stoves,
in later years. Bituminous coal ignited easily and burned with a vigor.
But many believed coal-fire fumes would contaminate food, or that
culinary perfection would be threatened by any other fuel than seasoned forest wood. There was evidence of this attitude in Cincinnati
where Pittsburgh coals had been introduced in the eighteen-twenties.
Advertisements for cook stoves usually advised that they could be
used with wood or coal, but often the listings noted only wood.[77]

Views of this kind did not prevail at Pittsburgh. By 1833 the
"smokey city" and its environs contained thirty thousand people. An
authoritative estimate placed the city's domestic coal consumption at
three and one-half million bushels, not including an additional million or more for stores and public buildings.[78] Twenty years later
Pittsburgh homes, stores, and public buildings used nearly eleven
million bushels of coal or more than three hundred and ninety-eight
thousand tons. This was one-third of the total consumption for the

---

[74] F. A. Michaux, "Travels West of the Allegheny Mountains, 1802," *Early Western
Travels 1748-1846*, ed. R. G. Thwaites (Cleveland, 1904-06), III, 153, 171.

[75] Fortescue Cuming, "Sketches of a Tour to the Western Country 1807-1809,"
Early Western Travels, IV, 77.

[76] J. N. Boucher, *History of Westmoreland County, Pennsylvania* (New York,
1906), I, 460.

[77] See *Cist's Weekly Advertiser*, 1846-1848.

[78] *Pennsylvania Senate Journal*, 1833-34, II, 482.

city.[79] These figures seem to prove that much of the Pennsylvania coal dug from its western beds, like the coal of the eastern part of the State, was used for no other purpose than simply to keep people warm and enable them to cook their meals.

Pittsburgh basked in the warmth of her coal fires. A little self-conscious over her sooty exterior and sulphurous air, she defended her appearance with medical opinion: "The smoke of bituminous coal is anti-miasmatic. It is sulphurous and anticeptic [*sic*], and hence it is perhaps, that no putrid disease has ever been known to spread in this place." ·Coal fires lighted early in the fall burned until late spring. It was claimed that there was less ague, bilious fever, dysentery, and yellow fever among the inhabitants whose bright hearths kept out damp and chill vapors thought to carry those afflictions.[80] The Board of Consulting Physicians, of Pittsburgh, reported in 1833 "that intermittent fevers, and diseases produced by malarial emanations, have never prevailed extensively in this city," and commented on the probability that the low death rate from cholera that year was due to the sulphurous air.[81] Pittsburgh's death toll from that dread disease had totaled only thirty-three that summer.[82]

Belief in the medicinal properties of the city's atmosphere was not reserved to Pittsburgh physicians. Eli Bowen, of Philadelphia, crown prince of the anthracite literati, visited Pittsburgh in the eighteen-fifties. He mourned, "But alas! for the smoke! There is too much of that here—our anthracital experience never could be overcome sufficiently to allow our bituminization." But even Bowen, whose lamentations continued for another paragraph, observed, "The sulphurous air, however, prevents eruptions of the skin, and people suffering with these disagreeable diseases, should forthwith eschew medicine, and take up a residence for a short time in Pittsburgh. It will cure 'em."[83]

A less sooty exterior was presented by the "Queen City of the West." Still, Cincinnati, which became a prime purchaser of coals from the Pittsburgh-Monongahela area, could not avoid completely the smudge and stain. Warmth and industrial energy were attended by the evils of excessive smoke. Cincinnati welcomed the appearance

---

[79] G. H. Thurston, "Pittsburgh As It Is," *The Pittsburgh Quarterly Trade Circular*, I, October, 1857, 29. (1 bushel equals 80 pounds; 28 bushels equals 1 ton).

[80] Samuel Jones, *Pittsburgh in the Year 1826* (Pittsburgh, 1826), pp. 31-32.

[81] Hazard, *Register of Pennsylvania*, XII (July-December, 1833), 46, 315.

[82] *Ibid.*, p. 46.

[83] Eli Bowen, *The Pictorial Sketch-Book of Pennsylvania* (Philadelphia, 1853), pp. 179-180.

of cleaner coals from Ohio and Virginia toward the end of our period, and often repeated that the "curse" of her "sister city" was the billowing smoke hiding friend from friend and mother from child.[84]

So extensive had become the consumption and waste of timber around the western river cities of the Ohio Valley, that by 1850 mineral coal was looked upon as a necessity of urban life.[85] The western Pennsylvania coal trade moved out of the rough confines of the Monongahela and Youghiogheny valleys, past the "golden triangle," and down the mighty Ohio to Cincinnati and Louisville. Part of the coal fleet continued to the Mississippi, the trunk of the huge western drainage basin. Some barges floated south to Memphis and New Orleans; others were towed north to the markets at St. Louis.

Cincinnati, Louisville, and smaller Ohio River towns had been introduced to Pittsburgh and Wheeling coals on a commercial scale by the eighteen-twenties. The fuel sold from twenty to twenty-five cents per bushel. At that time a cord of wood brought $2.50. This was a reasonable price, but coal still was the cheaper fuel.[86] Over the years the cost of wood mounted while the supply of coal increased and its price declined. By 1848 coal was retailing in Cincinnati for ten cents a bushel.[87] This does not mean that wood was eliminated from the domestic market. Woodyards remained in business and many who could afford wood preferred it in the kitchen fires. It has been pointed out, too, that manufacturers of kitchen stoves and even hot-air furnaces informed the public that their appliances burned both soft coal and wood.[88] At mid-century coal sold for one-quarter the cost of wood.[89]

Price and scarcity of forest fuel compared with cheapness and convenience of free-burning bituminous overcame most of the prejudice against the latter. A warm room and hot food were more important than a full wallet. In dismal contrast to Cincinnati's tubs of pure white lard, of which she was so proud, the sulphurous grime from thousands of chimneys poured into the air and settled in a brown mist over the "Porkopolis."

Cincinnati in 1840 consumed two and one-half million bushels of coal. A safe estimate would be that from one-fourth to one-third of that amount fed grate and stove. By 1858 the consumption had risen

---

[84] *Cist's Weekly Advertiser,* August 13, 1845, February 14, May 30, July 4, 1851.
[85] *Ibid.,* February 14, 1851.
[86] *Ibid.*
[87] *Ibid.,* May 9, 1848.
[88] *Ibid.,* February 29, 1848.
[89] *Ibid.,* February 14, 1848.

to fifteen million bushels or approximately half a million tons. Not all this fuel came from Pennsylvania mines. Pittsburgh, Monongahela, and Youghiogheny coals competed not only with Wheeling coal, but by this time were faced with new rivals. The challenge came from the cleaner, less sulphurous Pomeroy coal of Ohio and the Kentucky bituminous from the Peach Orchard mines along the Big Sandy.[90] Future events, however, would show that competition or no, Pennsylvania bituminous by 1860 still held the major markets of the Ohio Valley.

The winter of 1856-57 caught the nation in the grip of a cold wave which sent temperatures to record lows all over the country. Baltimore harbor was blocked with ice; Savannah registered eighteen degrees.[91] Freezing wind whipped down the valley of the Ohio and laid chill breath over Cincinnati and Louisville. Their residents shivered and suffered. Since spring adverse weather had prevented adequate supplies from reaching these markets. First flood, then prolonged drought lasting through the summer and fall of 1856, and finally the big freeze which halted lake and river traffic over the entire Midwest resulted in economic pressure and personal privation. Louisville gained some relief through small shipments of coal from Indiana via the Jeffersonville Railroad, but the amount was inadequate for her factories and sold for double its usual price to home owners. Dravo and Hyatt, Louisville coal merchants, were getting twenty-two cents per bushel in January, 1857.[92] Cincinnati, with a population of more than one hundred and ninety-six thousand, found wood and coal difficult to buy even after ration slips had been obtained from the emergency Coal Committee. Available amounts brought thirty cents a bushel and twice that on speculation.[93]

The rooms of the Coal Committee were mobbed by six to eight hundred freezing citizens, many of whom were women. Tables were smashed and the police, scratched, pushed, and bitten, were scarcely able to restore order. One German woman harangued the crowd and urged the ladies to revolt against the city government. Apparently no one listened to this "Red Republican" in all the excitement, except the reporter from the Democratic *Cincinnati Commercial*.[94]

---

[90] *Ibid.; Annual Statement of the Trade and Commerce of Cincinnati for the Commercial Year Ending August 31, 1858* (Cincinnati, 1859), p. 15.

[91] *Pittsburgh Gazette,* December 24, 1856, January 19, 1857.

[92] *Ibid.,* January 23, 1857.

[93] *Cincinnati Commercial,* January 23, 1857. Reprinted in the *Pittsburgh Gazette,* January 26, 1857.

[94] *Ibid.*

Pittsburgh sympathized with Cincinnati and called upon the charitable to remember the aid that city had extended to Pittsburgh after the great fire of 1845. The Cincinnati Relief Committee was organized and money was donated liberally to purchase and send coal by rail. "We who are in the midst of abundance should be the first to respond to their appeal," the *Gazette* reminded the soot-covered but warm people of Pittsburgh.[95] The sum of four thousand dollars was raised and fuel was sent by way of the Pittsburgh, Fort Wayne, and Chicago Railroad and its connections. The greatest difficulties were the shortage of cars and excessive freight rates. Unlike the famous Reading of eastern Pennsylvania, this railroad and its connecting lines were not equipped as coal carriers. River navigation had handled the coal traffic of the West since the inception of the trade, and the small amount of coal the spare rail cars carried barely relieved the sick and needy of Cincinnati. Pittsburgh estimated that thirty-two thousand families would require a minimum of thirty thousand bushels per day or two hundred and ten thousand each week. An average car hauled 250 bushels—that meant 120 cars per week drawn by ten locomotives, leaving every two hours.[96] The magnitude of this problem of supply and the expense defeated further efforts until thaw permitted relief by river. Pittsburgh did not miss this opportunity to repeat her plea to Cincinnati for cooperation in Congress for funds to improve the navigation of the Ohio by slackwater or canal.[97]

The suffering in the Ohio River towns is bleak evidence of the tremendous dependence urban populations of that valley placed on the coal supply from Pennsylvania.

---

[95] *Ibid.*
[96] *Ibid.,* February 4, 1857.
[97] *Ibid.*

CHAPTER II

Gas Light

EARLY EXPERIMENTS AND PROJECTS

D URING the period before the Civil War, material change in urban living had brought improvements in transportation, streets, and fire-fighting facilities. From the installation of central heating and plumbing in public buildings and private homes to the construction of waterworks, the metropolitan centers of the new nation threw off the attire of colonialism and clad themselves in the fashion of modern progress. Not the least of these civic improvements was gas light. The glow of whale oil lamps in street and dwelling was to become a memory in the large American cities by the end of the Civil War.[1] Illuminating gas, made from large quantities of bituminous coals and carried through miles of pipe to thousands of private and public lamps, transformed the dark, sleeping cities into islands of light and activity.

The British led the field in this significant improvement. While the dull glow of New Bedford whale oil cast long shadows in Philadelphia's streets and flickered in common room and kitchen of her inhabitants, London boasted of four gas companies manufacturing the new light from British coals and supplying over sixty-one thousand private and seventy-two hundred public lamps. By 1832 the London Gas Light and Coke Company, consuming twenty thousand chaldrons of coal annually, piped its product through 122 miles of main to furnish half of London's gas light.[2]

In 1798, William Murdoch, after successful experiments with illuminating gas in his native Cornwall, was commisioned to light the Soho Works of Boulton and Watt.[3] A few years later, in 1802, LeBon in Paris and Murdoch in London gave public demonstrations of gas

---

[1] Fredrika Bremer, *The Homes of the New World, Impressions of America* (New York, 1854), I, 255.

[2] William Matthews, *An Historical Sketch of the Original Progress of Gas-Lighting* (London, 1832), pp. 143-146.

[3] William Murdoch, "An Account of the Application of Gas from Coal to Economical Purposes," *Abstracts of the Papers Printed in the Philosophical Transactions of the Royal Society of London From 1800 to 1830* (London, 1832), I (1800-1814), 295.

lighting during the celebrations of the Peace of Amiens.[4] In that same year a young Englishman, Benjamin Henfrey, who had immigrated to Northumberland, Pennsylvania, patented a method of making gas light and demonstrated his "thermo-lamp" in Baltimore, Richmond, and Philadelphia. He proposed to illuminate these cities as well as all the Federal lighthouses.[5] LeBon had used wood; Murdoch, coal; Henfrey, coal and wood. Differing in the utilization of raw material, the experiments had one outcome in common: they were regarded as impractical and failed as commercial enterprises for lack of financial support.[6]

These demonstrations, nevertheless, did succeed in stimulating further laboratory experiments here and abroad. In the United States David Melville, of Newport, Rhode Island, produced "hydrogenous gas or inflammable air" from "pit coal" and burned it in a brilliant flame without smell or smoke.[7] Melville persevered with his gas apparatus for nearly seven years, but like Henfrey, was unable to convince anyone of its practicability. The distinction which Melville holds in the early American experimentation with illuminating gas is that he was the only one to manufacture gas directly and solely from coal.[8] European methods of utilizing coal were familiar to scientists of this country, but the great commercial value of this mineral, which abounded in the United States, seemed to be unknown to most Americans. Dr. Thomas Cooper, of Philadelphia, had complained of this in 1816 in the following words:

> Indeed, there is one reason for introducing gas lights here, which does not exist in England: in that country the precious article, coal, the foundation of all manufacture, is in universal use, and esteemed as it deserves. Here, we know not yet its value. We do not use it at all in the form of coak for our iron furnaces; we hardly know the use of it even for steam engines; it forces its way very slowly as fuel in our stoves and houses; we use none of the coal tar for our vessels; and in fact it is to the generality of our people a substance whose great value is experimentally unknown. Whatever tends to bring it into public estimation, will be a public benefit: for the seat of wealth and influence will ultimately

[4] G. T. Brown, *The Gas Light Company of Baltimore, A Study of Natural Monopoly* (Baltimore, 1936), pp. 10-11.

[5] Thomas Cooper, *Some Information Concerning Gas Lights* (Philadelphia, 1816), p. iii; *Eighth Census, Manufactures*, p. clxxii.

[6] Brown, *Gas Light Company of Baltimore*, p. 11.

[7] *Niles' Weekly Register*, VI (May 21, 1814), 198-199.

[8] Brown. *Gas Light Company of Baltimore*, p. 12.

be placed in every civilized country, there, where canals centre and coals abound.[9]

Baltimore claims the distinction of being the first American city to build a gasworks. Long familiar with James River and English bituminous, Baltimore, ignoring English methods, did not manufacture gas directly from coal, but from tar or pitch, the common distillate of pine knots. This procedure was patented in 1816 by Dr. Charles Kugler, a Philadelphia merchant and self-styled scientist. In this process, the melted pitch flowed into a red-hot retort and was reduced to its gaseous state. Carbon supplied the heat and hydrogen the illumination.[10] When used on a commercial scale the gas was piped from the retort to the storage tank or gasometer which was little more than a great cup inverted over a tank of water. From the gasometer, the product was drawn off into the mains and found its way to the consumer. This method was demonstrated on a small scale at Peale's Museum in Philadelphia. That same June in 1816, the Museum's famous collection of wonders moved to Baltimore under the direction of Rembrandt Peale, the son of Charles Willson Peale. Kugler's gas apparatus was offered again to the public, this time with spectacular success. Carbureted hydrogen gas from wood tar caught the imagination of some prominent Baltimore citizens, among them the editor of the *Federal Gazette,* William Gwynn. Both Rembrandt Peale and Gwynn gave the demonstration considerable publicity. Within a week a small group of local capitalists had been induced to form the Baltimore Gas Light Company. City Council was petitioned immediately for a franchise and a contract to light the city with gas.[11]

During the first few years of its existence, the Baltimore Gas Light Company fought a losing battle for capital and a more satisfactory method for making gas. Tar had proven too offensive and costly. The combined filth, odor, and pecuniary loss forced a reorganization of the company's stocks and physical plant. An English engineer designed new equipment and adopted bituminous coal as the raw material for the manufacture of the illuminant.[12] By the eighteen-thirties the company was doing a profitable business and consuming large amounts of bituminous coal.

---

[9] Cooper, *Gas Lights,* pp. vi-vii.

[10] *Ibid.,* p. 139.

[11] Brown, *Gas Light Company of Baltimore,* pp. 12-20.

[12] *First Annual Report of the Trustees of the Philadelphia Gas Works* (1836), Preface. Referred to hereafter as *Annual Report, PGW. Report to the Select and Common Councils of the City of Philadelphia by the Committee on Lighting the City with Gas* (1833), p. 5.

Philadelphia had watched the Baltimore experiment with interest and detached amusement. Not more than a year after the close of the War of 1812, a letter had been received by the city's government from a James M'Murtrie. Fresh from a sojourn in England, M'Murtrie claimed to have perfected an economical process for the manufacture of light from "stone coal gas." The City Councils appointed a joint committee of four to investigate "the effect and economy of gas-lights." The committee accomplished nothing and by 1819 perished in official idleness.[13] Philadelphia had resisted Henfrey, ignored Melville, rejected M'Murtrie, and scorned Kugler. The Committee of Councils, moreover, seemed oblivious to Dr. Thomas Cooper's famous observations of gas light which recommended Pittsburgh or Liverpool coal as the only raw materials for manufacturing pure, bright illuminating gas on a large scale.[14]

Meanwhile, New York and Boston, following Baltimore's example, had established gasworks. New York experimented with several materials and had decided on rosin, the distillate of turpentine. Boston used imported soft coals, chiefly from England. Rosin gas was added to equalize the illumination as different kinds of soft coals yielded varying qualities of gas.[15]

For many years Philadelphia blinked in the reflections of her northern and southern neighbors. Not until 1833 was another committee appointed by Councils. Almost immediately opposition threatened to defeat the renewed efforts. Some contended that explosions, fires, and loss of life would result from the use of dangerous "inflammable air." Other citizens, aroused over the prospects of a "constant digging in the streets," contamination of water mains, and the pollution of the rivers from the residue of the works, argued that "no reservoir will be able to contain the immense fetid drains from such an establishment, to the destruction of the immense shoals of shad, herring, and other fish with which they abound."[16] New Bedford whale oil interests baited the newly formed committee with an offer to sell oil to the city on a five-year contract for eighty cents a gallon, a price well below the market value. Public consumption of whale oil compared with private use was trifling. The Committee on Lighting the City with Gas interpreted the overture for what it

[13] *First Annual Report, PGW*, Preface.
[14] Cooper, *Gas Lights*, p. 15.
[15] *Report . . . on Lighting the City with Gas*, pp. 3-6.
[16] *Ibid.*, p. 32.

was—an effort to halt the construction of a gasworks—and promptly rejected the proposal.[17]

Protests of the Philadelphia citizenry were combated in a well-organized campaign. Fire insurance rates in New York, Boston, and Baltimore had not been raised for homes using the new illuminant. In fact, it was argued, gas light was far less dangerous than candles or oil lamps. Reports of explosions were scoffed at in testimonial letters from the mayors of New York and Boston. Philadelphians were invited to think hard on the great blessings of the new light in comfort, convenience, economy, and morality. For in the last instance, it was not overlooked that public gas lights would aid the City Watch on its nightly rounds. The advantages of a gasworks were crowned finally by the comforting opinion of leading medical men. Instead of becoming a menace to health, the fumes from the works "would have a tendency to correct or destroy atmospheric miasmata, which produces epidemic disease."[18] Since this included the dread cholera, the committee must have felt that with this final stroke it had adminis-istered the *coup de grâce* to further deprecation.

Bituminous coal was to be the source of Philadelphia's new light. A survey of the existing gasworks in the East convinced the committee that soft coal was the most economical material available.[19] One bushel of bituminous coal would yield one and a quarter bushels of coke, one quart of tar, and four gallons of ammoniacal liquor, all marketable commodities. This was not all. The coal was to be Pennsylvania bituminous. Philadelphia's example would institute a great demand for Pennsylvania gas coals, internal improvements would be expanded, and great profits would accrue to the Commonwealth and its citizens. The appearance of Pennsylvania economic sectionalism is not unusual, for this was an era of growing faith in the State's future as the fountainhead of the American economy.[20] Editorials and articles written by Pennsylvanians indicate this belief, and one can find it reflected also in reports by the General Assembly. In 1832 the Pennsylvania House of Representatives Committee on Agriculture digressed from its immediate province to say: "And it is to the minerals, and particularly to her coals, that this state is to be indebted for that pre-eminence in wealth, population and power, which is to

---

[17] *Ibid.,* p. 36.

[18] *Ibid.,* pp. 29-30, 44-47, 49-50, 53-58.

[19] *Ibid.,* p. 9. Cooper had said the same thing nearly twenty years before. Cooper, *Gas Lights,* p. vi.

[20] *Report . . . on Lighting the City with Gas,* pp. 9-10.

distinguish her future career."[21] This faith endured throughout the period, but it was twenty years before western Pennsylvania gas coals would find the means of cheap transport and supply the markets of the seaboard.

The Committee on Lighting the City with Gas also had been carried away by its own optimism regarding the quick, profitable sale of the by-products of distillation. Coal tar and ammoniacal liquor were regarded as nuisances in England and on the Continent until the sixties.[22] There was little demand for coal tar in Europe until chemists developed "aniline colours," of which Perkins' Mauve, discovered in 1856, was the first. The Glasgow Gas Works poured coal tar over coke for more rapid combustion under the retorts. In England small amounts were used to preserve timber, or naphtha was extracted from it and applied as a solvent for India rubber.[23] American shipbuilders experimented with coal tar in place of pine pitch, but found it unsatisfactory as a waterproofing material. The market was so limited that the Philadelphia Gas Works stored much of it in large tanks in its yard to await new applications for this annoying by-product.[24] Small amounts of ammoniacal liquor, or ammonia water, were purchased by chemists for practically nothing and used in the manufacture of ammonium chloride or sal ammoniac. There was some information regarding chemical fertilizer in the late forties and the liquor, when combined with sulphuric acid, yielded the compound ammonium sulphate. It was not until after the Civil War, however, that ammonium sulphate was applied in large amounts as fertilizer.[25] It was true that coke, the third product of distillation, did have a market, but even this was limited for a number of years. Gasworks burned much of it as fuel under the retorts to manufacture more gas, and the Philadelphia company noted that coke also was used to feed their office stoves and stationary steam engines. The rest was peddled in the open market at prices varying from six to twenty cents a bushel.[26] But coke competed with anthracite in the home and factory and it was not sold in large quantities until its use in the iron industry was established at the end of the period.

---

[21] *Pennsylvania House Journal,* 1832-33, II, 452.

[22] George Lunge, *Coal-Tar and Ammonia* (New York, 1916), III, 1046.

[23] *Ibid.,* I, 18-20.

[24] *Thirteenth Annual Report, PGW* (1846), p. 6.

[25] Lunge, *Coal-Tar and Ammonia,* III, 1045-1046.

[26] *Twenty-first Annual Report, PGW* (1856), p. 10.

The year 1833 was one of investigation and persuasion during which the idea of a gasworks was presented to the city. The following year the Councils appointed S. V. Merrick to visit London, the "Mecca" of gas light. A prominent engineer, Merrick, who later became president of the Pennsylvania Railroad, was instructed to tarry in Britain and then tour the Continent. Paris, Brussels, and Ghent were on his agenda. Merrick compared the plants of Britain, France, and the new Belgium. He discovered that the continental works were of English origin and under English control.[27] When he returned to Philadelphia to design and build the gasworks, Merrick corrected some of the construction errors he had observed in the foreign plants. Several years later, Merrick's successor, J. C. Cresson, visited England and reported, in a burst of civic ego and national pride, that the Philadelphia works were superior in engineering efficiency to those in Britain.[28]

Bituminous coal was used in the British works. Merrick confirmed the committee's view that this was the only feasible and economical raw material for the manufacture of illuminating gas. A true son of Pennsylvania, he reported to Councils "that every material used in the fabrication of gas, will be the product of Pennsylvania labor," from the bituminous coal from which it would be made to the anthracite for heating the retorts and the lime for purification. "And not a lamp will shed its rays over our streets," wrote Merrick, "which has not paid a tribute to the internal improvements of the state."[29] These bold, enthusiastic words indicate that Merrick was striving for effect and catering to Pennsylvania pride. A keen observer and competent engineer, he must have realized that anthracite would have been a costly fuel when the distillation of bituminous coal produced quantities of coke which could be used to heat the retorts in the common cycle of bituminous coal gas manufacture. Penned partially as propaganda, partially as prediction, the phrases were believed by many. The Philadelphia Gas Works was to be a proud monument to Pennsylvania resources and enterprise. Merrick, however, proved to be a false prophet. Although western Pennsylvania gas coals were given preference by the Philadelphia company whenever possible,[30] high freight rates over the Public Works prohibited free use until the late

---

[27] *Report . . . on Lighting the City with Gas,* pp. 58-60.
[28] J. C. Cresson, *Report to the Trustees of the Philadelphia Gas Works* (1845), p. 15.
[29] *Report . . . on Lighting the City with Gas,* p. 61.
[30] *Eleventh Annual Report, PGW* (1846), p. 10.

eighteen-fifties.[31] Meanwhile, Virginia coal and later English bituminous were used.[32] Cheap lime, derived from oyster shells raked from the beds of the Chesapeake or the shallow flats of the Delaware along the Jersey shore, furnished much of the purifying agent.[33] The fuel used to heat the gas coals was not Pennsylvania anthracite, but coke, the by-product of distillation of foreign or Virginia soft coals.[34]

Pennsylvania Gas Coals

During the first fourteen years of its existence the Philadelphia Gas Works operated without Pennsylvania bituminous gas coals.[35] The valuable resource from the western part of the State had not found its way across the Alleghenies to the eastern cities, and most of the coals used by the Philadelphia Works to 1848 came from the Chesterfield mines of Virginia. New York, abandoning rosin, relied upon British coals which came in ballast or by the keel as cargo equal to about twenty or twenty-four tons. The Brooklyn Gas Works by the mid-fifties was purchasing an estimated forty thousand tons of British coals per year.[36] New York supplemented her supplies with Virginia coals, and when the Deep River mines of North Carolina were opened, some of this coal was shipped by way of Wilmington to the New York market.[37] Wilmington, North Carolina, seemed to ignore the gas coals passing through her port and manufactured her illuminating gas from wood cut from the vast pine barrens of the Carolina coastal plain. The source was near and labor was reasonable. Wilmington, in the eighteen-fifties, gloried in the title of "the cheapest lighted city in the United States."[38] About this time, the *Miners' Journal,* commenting on the cost of generating illuminating gas from pine wood, noted that a Washington dentist had taken out a patent which would light all

[31] *Sixteenth Annual Report, PGW* (1851), p. 17.

[32] *Thirteenth Annual Report, PGW* (1848), p. 8; *Fourteenth Annual Report, PGW* (1849), p. 6.

[33] *Fifteenth Annual Report, PGW* (1850), p. 290; *Sixteenth Annual Report, PGW* (1851), p. 16.

[34] *Second Annual Report, PGW* (1837), pp. xix-xx; *Seventh Annual Report, PGW,* (1842), p. 4; *Eighth Annual Report, PGW* (1843), p. 4; *Ninth Annual Report, PGW* (1844), p. 217.

[35] *Sixteenth Annual Report, PGW* (1851), p. 17.

[36] CSHS, Palmer Papers, Notes: For the Executive Committee of the Westmoreland Coal Company, February 16, 1856.

[37] CSHS, Palmer Papers, H. Wavdell to W. J. Palmer, 1856.

[38] *Miners' Journal,* July 2, 1853.

the lamps in Norfolk for less than one dollar a night. This was to be commended, quipped the *Journal,* since it was even cheaper than moonshine.[39] Boston favored British and Nova Scotia coals, while some of the smaller eastern cities, among them Pottsville and York, Pennsylvania; Trenton, New Jersey; and Springfield, Massachusetts, used the more expensive rosin gas made from the distillate of turpentine.[40]

The Chesterfield mines produced the best available Virginia gas coal. Since it varied in quality and was exceedingly liable to spontaneous combusion, it was considered inferior to the western Pennsylvania fuel.[41] A shortage of Chesterfield coal in 1848 forced the Philadelphia company to buy large quantities of the higher-priced British bituminous. British Newcastle yielded nearly one-third more gas than Chesterfield, but the illuminative power was weak. Later a blend of Virginia, British bituminous, and cannel coals was tried with some success.[42]

At mid-century the Pennsylvania Canal Commission reduced the rates on the Public Works and Pittsburgh gas coals were shipped to Philadelphia for the first time in large quantities. More than half the coal used during 1850 came from the Pittsburgh beds. The officials of the gasworks thought their problems had been solved. Here was a coal of tried reputation. Western cities had used it for years and always had reported satisfactory results. The native gas coal was put to use and carefully watched. Its high illumination eliminated the expensive addition of resinous material, but the company's engineer expressed disappointment in its other characteristics. The coke was not as free-burning nor the gas yield as large in volume as the British coals. Even so, it was the best coal used to that time, and the company deemed it prudent to secure as much as possible while the price remained reasonable.[43]

The following year, increased freight rates on the Pennsylvania Canal again interfered with the shipments of Pittsburgh coal eastward. The Philadelphia company reported to its stockholders, "The current trade on the State Canals not being favorable for obtaining large supplies of coals from Pittsburgh, very little has been derived

[39] *Ibid.,* March 5, 1853.
[40] *Ibid.,* March 16, 1850.
[41] *Eleventh Annual Report, PGW* (1846) , p. 10; *Fourteenth Annual Report, PGW* (1849) , p. 8.
[42] *Fourteenth Annual Report, PGW* (1849) , p. 9.
[43] *Sixteenth Annual Report, PGW* (1851) , p. 17.

from that quarter." A small amount was obtained, but Virginia and English coals once more made up the bulk of the gasworks' purchases.[44] The company at this point seems to have abandoned hope that the superior Pennsylvania coal ever would become available in reasonable and dependable supply.

The growth of the Philadelphia company and consequently the consumption of coal had been little less than amazing. In operation only four years by 1840, the works supplied twenty thousand dwellings and seven hundred public buildings.[45] This was accomplished through the carbonization of less than five thousand tons of coal. By 1852 the Philadelphia Gas Works had become one of the largest, if not the largest, in the country, consuming over twenty thousand tons of coal annually.[46] Four years later this amount had more than doubled and was to redouble again before the end of the Civil War.[47] Twenty-five thousand tons supplied illumination for over two hundred thousand lamps in 1853. A decade later more than eighty thousand tons of coal were carbonized, and gas flowed through 451 miles of main to furnish illumination for half a million lamps.[48]

The price of gas to the consumer wavered between two and three dollars per thousand cubic feet.[49] It would be unrealistic to suggest that the company, controled by the city of Philadelphia since 1841, was not making a profit. Nevertheless, during the fifties the officials complained of rising operating costs. Between 1852 and 1854 labor had increased its demands eighteen per cent and bituminous rose in price thirty-five per cent.[50] Ammoniacal liquor and coal tar often went begging, whereas coke in Philadelphia and in New York, competing with the popular anthracite coal, found a slow market.[51] The Philadelphia concern felt the rate pressures from the State Works and returned to the less efficient Virginia and more costly British coals. The latter soared to unprecedented heights during the Crimean War and the eastern gasworks complained of the exorbitant British freight rates on coal. New western Pennsylvania gas coal from Westmoreland

---

[44] *Seventeenth Annual Report, PGW* (1852), p. 345.
[45] *Journal of the Franklin Institute,* XXXI (January-June, 1841), 231-241.
[46] *Ibid.,* LV (January-June, 1853), 207.
[47] See Table B, Appendix.
[48] *Twenty-ninth Annual Report, PGW* (1863), p. 16; *Thirtieth Annual Report, PGW* (1846), p. 16.
[49] See Sixteenth through Twenty-sixth *Annual Report, PGW* (1851-1861).
[50] *Twentieth Annual Report, PGW* (1855), p. 24.
[51] *Seventeenth Annual Report, PGW* (1852), p. 347.

County was greeted favorably, particularly in New York, provided it could be brought east by rail at reasonable rates.[52]

Increased costs forced the Philadelphia firm to intensify its campaign against destructive spontaneous combusion which volatilized the coal stockpiles. At the same time the company began conducting experiments in efficiency to obtain the largest yield of gas and the greatest intensity of illumination with the least amount of raw material and labor.[53] Some consideration was given to a revived patent which produced "water gas" by applying steam to hydrocarbonaceous material. Dr. Abraham Gesner's discovery which generated illuminating gas from the asphaltum of New Brunswick province of Canada was listed as an alternative. A French experiment with pure hydrogen and a platinum wick also was noted as a possibility. The company's engineer defended this novelty by remarking that it was not as "absurd and impractical" as the projects a year or two before concerning electric light.[54]

Hard pressed during the Crimean War, which placed it at the mercy of British carrying charges, uncertain of western supplies, and disgusted with the low quality of gas coal from the South, Philadelphia seriously considered turpentine rosin or wood when obtainable at low price. A special apparatus was constructed from a "foreign patent." Wood gas was quite satisfactory, and it was found that under the control of the new retort, one cord of ordinary firewood gave off nearly twice as much gas as a ton of the best Pittsburgh coal.[55] The manufacture of wood gas at first was used to resist further increase in the price of coal.[56] But wood gas experiments continued and ten additional retorts were put into use in March, 1855.[57] By 1857 the engineer reported,

> With the present relative prices of wood and coal in the Philadelphia market, the cost of making gas from the former is somewhat lower, but the difference is not sufficient to justify the immediate abandonment of the latter. Should a commercial change occur, by which the price of coals should

[52] CSHS, Palmer Papers, Notes: For the Executive Committee of the Westmoreland Coal Company, 1856.

[53] *Seventeenth Annual Report, PGW* (1852), p. 347.

[54] *Ibid.*, p. 349; *New York Journal of Commerce.* Reprinted in *Cist's Weekly Advertiser*, February 13, 1850.

[55] *Twentieth Annual Report, PGW* (1855), pp. 4-23.

[56] *Ibid.*, pp. 23-24.

[57] *Twenty-first Annual Report, PGW* (1856), pp. 17-18.

be again advanced to the high point reached two or three years ago, there might arise important advantages to these works and its customers, from the ability to make the substitution of wood for coal.[58]

Scientists in Philadelphia and New York had analyzed wood gas and had found its illuminating qualities equal to coal gas. The find- ings were not surprising, for the Philadelphia Gas Works was well aware of the three major methods of gas manufacture. The first was from bituminous coal, employed by most large cities in the United States and abroad; the second was from wood gas, used by several of the southern cities and by some towns in Europe; the third was the "water-gas" method, in which steam was played upon carbonaceous material, especially turpentine rosin. This latter method had been used in practical tests by the Northern Liberties Works of Philadel- phia for a few weeks in 1860.[59] After years of experimentation it was decided at last that bituminous coal was still the most economical material to be used. Rosin cost no less and "water gas" would need new equipment. Supplies of reasonably priced pine wood were even less reliable than bituminous coal, as wood costs chiefly depended upon wages.[60]

A basic force in the decision to continue to use bituminous was the increased supply of Pennsylvania western coals moving east over the Pennsylvania Railroad system. The city of Philadelphia, owner of the gasworks, was a large stockholder in the line; thus, freight rates could be discussed thoroughly. Despite experimentation and investigation of other raw materials, western Pennsylvania bituminous, by 1858, had not only eliminated wood and "water gas," but had replaced Virginia and British coals in Philadelphia.[61]

The Westmoreland Company, one of the great gas coal concerns, moved into the eastern market in 1855-56.[62] On the eve of the Civil War this concern supplied fifty-eight gasworks. Purposely seeking the eastern marts, it sent more coal to Philadelphia and New York than to Pittsburgh and the western cities.[63]

With the beginning of the Civil War, Philadelphia made hasty pur- chases of foreign coal from the port of New York, since the city feared

---

[58] *Twenty-second Annual Report, PGW* (1857), p. 17.

[59] *Twenty-sixth Annual Report, PGW* (1861), p. 18.

[60] *Ibid.*, p. 19.

[61] W. J. Nicolls, *The Story of American Coals* (Philadelphia, 1897), p. 354.

[62] CSHS, Palmer Papers, Coal Notes and Pocket Diary, 1856-1857.

[63] *Twenty-seventh Annual Report of the Philadelphia Board of Trade* (1860), pp. 117-118.

the curtailment of supplies caused by military demands on rail transportation.[64] The fears were unfounded, as western Pennsylvania gas coals continued to be moved into Philadelphia during the Rebellion. An inflated currency forced coal prices upward. In 1863-64 domestic bituminous was bringing $11.00 to $11.40 per ton, twice its usual cost.[65] The consumption of the Philadelphia works in 1864 had grown to ninety thousand tons, but even this amount, chiefly shipped from the Pennsylvania Gas Coal Company and Westmoreland mines, did not satisfy demand.[66]

The story of the triumph of Pennsylvania gas coals in the eastern market is further evidence that transportation and utility do not exist in separate economic vacuums. The account of the Philadelphia Gas Works, while not intended to be interpreted as a conclusive example of utilization and experimentation, can, nevertheless, be employed as a model to show how American enterprise applied a basic resource to the material progress of the nation. In Pennsylvania the decade of the fifties was marked by an increasing number of requests for incorporation of gasworks.[67] By 1862 the United States counted 420 gas companies with a total capital investment of more than five million dollars.[68]

It is obvious from the preceding pages that although these companies burned large amounts of bituminous coals, not all their fuel came from the mines of Pennsylvania, nor has it been possible to determine exactly how much Pennsylvania gas coal was used by these works. On the other hand, it would not be an error to state that by 1860 Pennsylvania gas coals had begun to supplant foreign coals in the East, and were firmly established in the Ohio and Mississippi valleys as well as in the Great Lakes region.

Pennsylvania gas coals found some competition in the West from Ohio, Indiana, and western Virginia bituminous, but as a source for illumination they were in great demand. During the drought of 1854, St. Louis found her stocks of Pittsburgh coal rapidly diminishing. There was little prospect of fresh supplies by water, and for a time St. Louis was plunged into darkness.[69] The St. Louis Gas Light Com-

---

[64] *Twenty-eighth Annual Report, PGW* (1862), p. 18.

[65] *Thirtieth Annual Report, PGW* (1864), p. 4; *Thirty-first Annual Report, PGW* (January, 1865), p. 7.

[66] *Thirty-first Annual Report, PGW* (1865), p. 7.

[67] See *Pennsylvania Senate Journal*, 1852 to 1855 and 1857 to 1860-61, I.

[68] *Eighth Census, Manufactures*, p. clxxii.

[69] *Pittsburgh Gazette*, January 28, 1854.

pany purchased approximately fifteen to eighteen thousand tons of Pittsburgh gas coals annually.[70] The city of Pittsburgh maintained three gasworks, which had a combined yearly consumption totaling nearly 12,500 tons.[71] From Cincinnati to Louisville, south to Memphis and New Orleans, and north to Chicago, Cleveland, and Erie, Pittsburgh gas coals found ready sale.

Pennsylvania bituminous coal, black brand of light, entered the urban centers of the United States and contributed to the welfare and progress of a youthful, energetic nation which, in its quest for convenience, comfort, and material gain, successfully utilized one of the great treasures of nature.

---

[70] *Ibid.,* July 29, 1859.
[71] *The Pittsburgh Quarterly Trade Circular,* I, 29.

CHAPTER III

Mineral Fuel and
Industrial Growth

COAL, STEAM, AND MANUFACTURING—THE WEST

IN THE industrial civilization of the nineteenth century, steam was the giant of progress, the sinew of national strength, and the raw force in the manufacturing supremacy of Western man.[1]

In America, with the exception of early isolated experiments, steam was produced first by wood and charcoal. Later, along the eastern seaboard, Virginia and English soft coals comprised part of the fuel. At Pittsburgh and in sections of the western country where bituminous was abundant, this mineral fuel was employed very early in the industrial use of steam power. Steam drove the spindles and shuttles of the cotton factories, the saws in the lumber mills, and provided power for rolling works and forges. In numerous ways the stationary steam engine, burning bituminous coal, gave its energy to the industrial development of the western river valleys.

Echoing a phrase from Hamilton's *Report on Manufactures*, Tench Coxe, in 1794, re-emphasized the significance and prevalence of the "useful fossil," coal, in the "middle and western country" and the Forks of the Ohio. His statements, intended for the immediate times, foretold the future of Pennsylvania, and especially of Pittsburgh, as cradles of industrial development through the use of mineral fuel.[2]

Located in the depression of the converging river valleys, young Pittsburgh presented a drab appearance. A cloud of brown coal smoke hung over the town. Unpaved, crooked streets, where hogs, dogs, and people sloughed in mud, were bounded by grime-soaked brick and wooden structures. All was hustle, bustle, and filth. From the cursing boatmen at the river wharves to the scheming shopkeepers preying upon the constant flow of human traffic westward, an aura of mate-

---

[1] "And who will write the biography of Steam itself—that giant of the nineteenth century— . . . ?", *New York Times*. Reprinted in the *Miners' Journal*, June 4, 1853.

[2] Tench Coxe, *A View of the United States of America* (Philadelphia, 1794), p. 70.

41

rialism prevailed.[3] During the first three decades of the nineteenth century, blacksmith shops, brickyards, glass factories, and iron and textile mills using coal sounded a hum of industry which someday would grow into a roar to be heard in every part of the Union. Abundant, cheap, easily accessible, and readily combustible bituminous fuel prophesied the day of the Titan to be close by.

Location alone would have made Pittsburgh a significant city, for standing where the Allegheny and the Monongahela rivers meet to create the Ohio, she formed the point of departure for the westward movement into the Ohio and Mississippi valleys. But Pittsburgh had more to her credit than geographic location. She struck her roots deep into one of the greatest coal beds of the world. She attracted the raw iron from the furnaces of the surrounding ore-rich counties of Somerset, Westmoreland, and Fayette. With her plentiful mineral fuel Pittsburgh rolled and forged the pigs and blooms into nails, tools, and agricultural implements. These products supplied the needs of transients who would soon form the permanent population of the West and remain consumers of Pittsburgh's goods.[4] Cheap fuel was of incalculable importance in the growth of other industries, especially in glass and textiles. In the numerous mills of the city steam engines consumed thousands of bushels of coal.[5] Conscious of her natural advantages in geography and resources, Pittsburgh encouraged her factories and attracted skilled, energetic people and capital investment.[6]

The early manufacturing establishments of the growing city used coal as their principal fuel. Breweries, glass works, foundries, and machine shops were erected near easily accessible diggings. Some were supplied by wagons or even by sleds operating on gravity runs "greased" with mud, while large quantities of coal also were ferried

---

[3] Jones, *Pittburgh in the Year 1826*, p. 36; Thomas Nuttall, "A Journal of Travels into the Arkansas Territory, During the Year 1819," *Early Western Travels*, XIII, 45; Eastwick Evans, "A Pedestrious Tour, etc.," *Early Western Travels*, VIII, 249-260; Una Pope-Hennessy (ed.), *The Aristocratic Journey* (New York, 1931), pp. 288-289.

[4] C. E. Reiser, *Pittsburgh's Commercial Development, 1800-1850* (Harrisburg, 1951), pp. 4-9.

[5] Amount of Bituminous Coal Consumed in Manufacturing at Pittsburgh—

1825—	1,000,000 bushels	1838—	357,140 tons
1833—	245,910 tons	1842—	420,000 tons
	1846—	678,572 tons	

Quoted from *Ninth Annual Report of the President and Directors to the Stockholders of the Cleveland and Pittsburgh Railroad, 1856*, p. 27.

[6] Reiser, *Pittsburgh's Commercial Development*, pp. 4-9.

from Coal Hill across the river to supply Pittsburgh proper.[7] Little experimentation with fuel was necessary. With few exceptions, equipment or processes utilizing wood could be shifted to bituminous coal which ignited easily and burned with a will.

One of the first American pioneers in steam power, Oliver Evans, who had moved from Philadelphia to the fringes of the frontier at Pittsburgh, appreciated the utility and efficiency of coal in steam engines. Evans helped to introduce to Pittsburgh the high-pressure steam engine for ship and mill. In 1817 Pittsburgh boasted of three steam-engine plants turning out equipment for river boats and factories. The city also counted two steam gristmills, a nailery, woolen mill, paper factory, and sawmill, all powered by steam generated through the heat of bituminous coal. In Brownsville, a cotton carding and spinning factory added to the growing enterprise of steam and coal.[8]

The use of steam and coal expanded steadily as the city increased in population and the West filled up with homesteads and towns. Not without reason was coal looked upon as "the pedestal" of the "manufacturing emporium" of western Pennsylvania by the Pennsylvania Senate committee investigating the coal trade in 1833-34. Pittsburgh and its environs possessed ninety steam engines by 1833, consuming annually an estimated sixty-nine thousand tons of coal.[9] Bituminous coal played its persistent part in the manufacture of Pittsburgh white lead, glass, edge tools, tanned leather, flour, paper, and textiles, not to mention the place it held in the iron industry. From pottery to printing, the fuel was called upon to furnish heat or create steam.

Even Pittsburgh distilleries found coal indispensable. If colonial New England had "floated on a sea of rum," Pittsburgh waded in rivers of whiskey. It would not be unlikely to suppose that considerable coal dust coated the jugs containing Sutton's famous "Tuscaloosa," drunk from Maine to Georgia, or that the popular "Pure Rock Water" held a peculiar bituminous flavor as it, to paraphrase the words of a contemporary, stole gently upon the senses and animated the intellect without ever collapsing an idea.[10]

In the long view of fifty years of Pittsburgh's industrial enterprise through coal, one obvious fact persists: the raw materials which fed

---

[7] J. S. Wall, *Report on the Coal Mines of the Monongahela River Region, . . . Second Geological Survey of Pennsylvania, Report of Progress K4* (Harrisburg, 1884) , p. K4 xxi.

[8] *Ibid.,* p. K4 xxiii.

[9] *Pennsylvania Senate Journal,* 1833-34, II, 483.

[10] Jones, *Pittsburgh,* p. 80.

her mills and shops came to the source of cheap fuel.[11] Iron from the furnaces of western Pennsylvania, copper from Michigan, lead from Illinois, cotton from Tennessee, and even supplies of sand from as far away as Cape Girardeau on the Mississippi were transported to "the city of the three rivers."[12] Pittsburgh coal turned cotton into cloth, sand into glass, and the cumbersome lumps of iron into a thousand useful items. Coal was extracted so near to the mills that as late as 1857 it could be purchased for as little as $1.20 to $1.50 a ton, or from four to five cents a bushel.[13] This price had not been altered to any extent in half a century. Some factories owned their mines and reckoned the cost of a ton at less than one dollar.[14]

Downriver towns such as Cincinnati and Louisville found coal ranging up to two dollars a ton more than at Pittsburgh. During the coal famine of 1854-55 and again in 1856-57, short supply pushed costs to nine dollars and even fifteen dollars a ton. Factories either bought at these prices or closed their doors.[15]

In freezing weather, spring freshet, or extended drought, Pittsburgh's coal supply was readily available and remained steady in price. Although adverse weather conditions could slow the transportation of other raw materials and finished products, the city held a distinct advantage over most centers of manufacturing in the West. In 1857 the *Pittsburgh Gazette,* remarking on the coal shortage in Cincinnati, Louisville, and the small Ohio River towns, said, "If they cannot find some access to a constant supply of cheap fuel, those who are seeking location of workshops and factories will pass by them, whatever other advantages they possess, to those places that have the source of all industrial power."[16] Pittsburgh manufacturers felt secure in their citadel of coal as they used bituminous fuel for steam, iron, and glass, the largest industrial consumers of their treasured resource.[17]

---

[11] *Pennsylvaina Senate Journal,* 1833-34, II, 483.

[12] *Cannelton, Perry County, Indiana . . . Its Natural Advantages as a Site for Manufacturing* (Louisville, 1850), pp. 31, 38.

[13] *The Pittsburgh Quarterly Trade Circular,* I, 23.

[14] *Ibid.*

[15] *Ibid.*

[16] *Pittsburgh Gazette,* January 7, 1857.

[17] A rough estimate of Pittsburgh's annual coal consumption was attempted in 1854 by William J. Tenney in *The Mining Magazine,* II (January-June 1854), 213-216.

Glass Factories (20) —	600,000 bushels
Cotton Mills (5) —	100,000 bushels
Rolling Mills (17) —	6,500,000 bushels
Foundries, machine shops, steamboats, and public buildings—	22,305,000 bushels

Although the city was proud of her collieries, Pittsburgh's industrial strength did not rest upon coal alone. Standing at the gateway to the western country, she looked across three rivers to their valleys beyond. Improvements of these natural waterways in the forties and fifties increased her advantages in trade. When the high freight rates and inadequacies of the Pennsylvania State System became evident, Pittsburgh citizens campaigned for rail connections with the East. The Pennsylvania Railroad was joined to the city in 1852 and carried the manufactured products and coal to the seaboard. The Cleveland, Connellsville, and the Pittsburgh, Fort Wayne and Chicago lines added spokes to the wheel of commerce of which Pittsburgh was the hub. By the Civil War the combination of location, transportation, and the use of coal as a source of industrial energy had completed the economic cycle of the city.

The coal trade of the western part of the State increased in the forties and fifties. Pennsylvania bituminous was brought down the Ohio or carried north to the Lakes, while large amounts were hauled over the Pennsylvania Railroad to the eastern areas. Much of this fuel was used in the industrial plants of the urban centers. Factories at Cincinnati, Louisville, and St. Louis were important consumers. Monongahela coal was used to make steam in New Orleans, and Cleveland began buying Pennsylvania coals in quantity by the mid-fifties.

Cincinnati acknowledged Pittsburgh's supremacy in iron, glass, and cotton manufacture, but the Ohio city emphasized that in 1840 she consumed two and one-half million bushels of coal annually, in 1845 nearly five million, and by 1850 used fifteen million bushels. The major cause of this large increase in coal consumption was the growth of manufacturing during those ten years.[18] Lumber mills, sash factories, flour mills, lard-rendering machines, and a new waterworks, all powered by steam using bituminous coal, were to be found in the "Queen City of the West."

Thus Cincinnati, like Pittsburgh, looked to coal for her source of industrial power. Until the year 1840 the fuel came from the Pittsburgh and Wheeling areas. Ohio coal from the Pomeroy mines and Peach Orchard from Kentucky began competing with the Pennsylvania product in the late forties.[19] The Ohio and Kentucky coals were prized for their cleanliness, but the mines of these regions were un-

---

[18] *Cist's Weekly Advertiser*, August 13, 20, 1845, April 24, 1850.
[19] *Ibid.*, February 14, 1851.

able to furnish enough fuel for Cincinnati's homes and factories. Pittsburgh, Monongahela, and Youghiogheny coals continued to dominate the markets of Cincinnati and the rest of the Ohio Valley to 1860.

While Pennsylvania bituminous was supplying the mills of town and city, other demands were being made upon it by an old and important industry. Saltworks were the first large consumers of coal in Pennsylvania. A basic commodity in an agrarian economy, salt was in constant demand and had no difficulty in finding a market. The earliest saltworks in Pennsylvania was opened on Big Beaver Creek in 1784. About 1801 bores were made in the banks of the Conemaugh and brine was struck at a depth of 450 feet.[20] When the Onondaga salt supply was cut off during the War of 1812, western Pennsylvania, long dependent on New York salt, quickened her search for local sources.[21] The search proved worthwhile. Within a decade there were twenty works along the Kiskiminetas sending twenty thousand barrels of salt to Pittsburgh each year.[22] Twice as many works were in building, and so significant had become the new industry that one of the major arguments for the "golden link to the west," the Pennsylvania Canal, was the desire to bring the salt trade east of the Susquehanna.[23] By 1826 there were thirty-five works on the Conemaugh and Kiskiminetas.[24] Salt wells continued to be brought in along the banks of other streams and rivers in the heart of the western Pennsylvania coal country. Crooked Creek, the Chenango, Saw Mill Run, the Mahoning, and Big Sewickley, as well as major water lanes, the Allegheny, Youghiogheny, and Monongahela, added to the lists of works which numbered more than ninety by the mid-thirties.[25] These works produced thousands of bushels of salt annually, and each year consumed large amounts of bituminous coal in their steam pumps and under their salt pans. It is difficult to estimate accurately the amount of mineral fuel consumed by these works as the coal was mined locally and complete records were evidently not kept. In 1834 Pennsylvania official reports stated that five million bushels of coal were burned in the production of one million bushels of salt.[26] The

[20] *Eighth Census, Manufactures*, p. ccii.

[21] N. B. Craig, *History of Pittsburgh* (Pittsburgh, 1851), p. 284.

[22] *Pennsylvania House Journal*, 1824-25, II, 269.

[23] *Ibid.*

[24] *Eighth Census, Manufactures*, p. ccii.

[25] *Pennsylvania Senate Journal*, 1833-34, II, 482-83.

[26] *Ibid.*

federal census of 1860, in noting the amount of salt produced in Pennsylvania in 1840 and 1850, does not mention how much bituminous coal was used by the works.[27]

The consumption of Pennsylvania coal in salt production was not limited to the saltworks within the confines of the State. The four large salt districts of New York—Syracuse, Salina, Liverpool, and Geddes—gained access to the semibituminous coal of the Blossburg fields in the forties, and depended on this supply of fuel to feed their numerous evaporating blocks.[28] The Blossburg region, a small isolated sector east of the Great Bituminous Coal Field of the Alleghenies and more specifically near the headwaters of the Tioga River in north central Pennsylvania, found its market in New York State. A railroad about forty miles in length conveyed thirty thousand tons a year to the Chemung Canal at Corning, New York. The coal then was transported over New York canals to the salt centers.[29] The significance of the trade was appreciated by the two states. Governor Seward of New York commented favorably on the flourishing trade between his state and Pennsylvania via the Chemung Canal.[30] New York permitted all coal destined for the state-controlled saltworks to pass toll free on her canals.[31] This was quite different from the policy of the Pennsylvania System, or for that matter from that of Pennsylvania herself. Often the tolls were so high as to prohibit the movement of Pennsylvania coals to the eastern markets, and more than once politicians had proposed a tax on every ton of coal mined in the State.[32]

Fuel supply for the New York saltworks was considered a choice plum in the Pennsylvania coal trade. It was estimated that one ton of coal was needed to evaporate forty-five bushels of salt in the New York works, or twenty-five to thirty bushels of coal for every forty-five bushels of salt.[33] This did not take into consideration the coal used to fuel the steam pumps. During the late forties, when the salt makers conducted experiments with anthracite, the Delaware and Hudson Company, which supplied much of the Hudson Valley, tried to capture this lucrative business. The board of directors voted to appro-

---

[27] *Eighth Census, Manufactures,* p. ccii.

[28] *Ibid.,* p. cci.

[29] R. C. Taylor, *Statistics of Coal* (Philadelphia, 1855), p. 341.

[30] Samuel Hazard, *United States Commercial and Statistical Register,* VI, January 19, 1842, p. 37.

[31] Taylor, *Statistics of Coal,* p. 341.

[32] *Pennsylvania Senate Journal,* 1845, II, 18; *Miners' Journal,* September 13, 1851.

[33] *Eighth Census, Manufactures,* p. ccii.

priate not less than $10,000 to be spent by its coal committee for the sole purpose of promoting the use of anthracite in salt manufacture at Syracuse and Salina.[34] New York, however, continued to consume the cheaper and more accessible Blossburg semibituminous. By 1854 three hundred tons daily were shipped along the route of the Blossburg and Corning Railroad, and most of it found ready use in the saltworks of Pennsylvania's northern neighbor.[35]

STEAM, COAL, AND MANUFACTURING—THE EAST

East of the mountains, anthracite coal had found a rich market through industrial use. The place which it held in manufacturing had not been attained with the ease surrounding the successful application and utilization of the free-burning Pennsylvania bituminous. Anthracite for industrial enterprise, like anthracite for domestic use, required special equipment for peak efficiency. Grates, fireboxes, boilers, and flues had to be built or altered to suit the fuel. What was even more important, the hard coal from Pennsylvania, whether it was to be employed in producing steam for stationary engines or in supplying heat for brick kilns, had to be sold to manufacturers who harbored no little prejudice against it.

An example of early disdain for the industrial use of anthracite can be observed in the action of the Senate of Pennsylvania in 1814. The Mutual Assistance Coal Company of Philadelphia presented a memorial to the Pennsylvania Senate calling attention to the importance of facilitating the introduction of "stone coal" into the city and its vicinity for use in the various branches of manufacturing. The memorial specifically petitioned the Senate to consider measures which would improve roads and the inland navigation of the Delaware, Schuylkill, and Lehigh so that anthracite could be brought to tidewater at reasonable prices. It was referred to the Committee on Roads and Canals.[36] Three weeks later the memorial was read out of committee with the recommendation that it no longer be considered.[37]

In 1815 the Mutual Assistance Coal Company of Philadelphia, in an attempt to introduce Schuylkill coal, printed testimonials on the uses of anthracite in manufacturing.[38] During the War of 1812 a

---

[34] Delaware and Hudson Minute Books, October 11, 1848.

[35] Taylor, *Statistics of Coal*, p. 341.

[36] *Pennsylvania Senate Journal*, XXIV, 1813-14, p. 161.

[37] *Ibid.*, p. 447.

[38] Scharf and Westcott, *History of Philadelphia*, I, 582.

Portsmouth, Rhode Island, anthracite firm had issued a pamphlet in the same vein.[39] When the second decade brought about the real beginnings of the anthracite trade, similar testimonials again appeared in pamphlet form. Brewers, blacksmiths, gunsmiths, and ironmasters certified the practical uses of anthracite in their occupations.[40] Although part of a promotional campaign, the testimonials indicate that the fuel was undergoing gradual acceptance outside its native areas. But it was some years before experiments and technological improvements would permit anthracite to gain a wide following in manufacturing and steam.

The first successful practical application of anthracite coal for generating steam was in 1825 at the Phoenix Nail Works of Jonah and George Thompson on French Creek, Chester County, Pennsylvania.[41] The works were constructed with the view that anthracite would be the fuel for its stationary steam engines. Since the flames of anthracite coal, unlike those of wood or bituminous coal, did not extend far enough from the grate to generate steam, it was decided that either the boilers and flues had to be made shorter than usual, or, as in the case of the Phoenixville plant, horizontal cylinder boilers should be used with a firebox and grate placed at either end.[42] Joshua Malin, an early anthracite convert, played a large part in designing the Phoenix Nail Works. He also was the first Philadelphian to use anthracite coal in the iron industry and claimed "without any fear of contradiction to be the first man who melted pig iron with Anthracite coal in this country."[43] As a steam engineer and ironmaster, Malin continued to erect steam engines throughout the State, both east and west of the mountains. He was convinced of the superiority of anthracite as a steam coal, and in a letter to the Franklin Institute in 1827 stated in positive terms that he knew of "no fuel that will generate steam so fast as the anthracite coal, when properly applied."[44]

Improvements in anthracite stationary steam engines continued for many years and, like improvements in heating and cooking apparatus,

[39] *Observations on the Rhode Island Coal* (1814).

[40] *Facts Illustrative of . . . Lehigh Coal.*

[41] *American Journal of Geology and Natural Science.* Reprinted in the *Miners' Journal,* August 20, 1831. *Eighth Census, Manufactures,* p. clxxi.

[42] *Journal of the Franklin Institute,* III, 191.

[43] HSP, Joshua Malin to Gerard Ralston, esq., April 20, 1827.

[44] *Journal of the Franklin Institute,* III, 191.

were encouraged by learned societies such as the Franklin Institute.[45] Marcus Bull's laboratory experiments had proven the heating power of anthracite for manufacturing, home heating, and for making steam. Bull, though chided by Boston scientists, was praised highly by James Renwick, one of the foremost steam engineers of the day. Renwick concurred with him and with Malin that anthracite, properly applied, could generate steam more rapidly than bituminous coal or wood because of the intense heat it threw off in the process of combustion.[46]

Two rather significant improvements on the anthracite stationary steam engine were patented in 1828 and 1829. The first, invented by Benjamin B. Howell, was given considerable attention by the Franklin Institute. This in itself was no mean achievement, as the society could be particularly caustic when reviewing an impractical patent. In commenting on Howell's invention, the editor of the *Journal of the Franklin Institute* made some interesting statements pertaining to the popular use of anthracite: "After the coals are once placed in our grates, and ignited, their motto seems to be 'Laissez nous faire,' and observing this, every thing proceeds with the utmost facility." He went on to say that anthracite first triumphed in the parlor, then in the kitchen, but "the fireman of the steam engine, and the ironmaster however, yet remain unconvinced." Anthracite would triumph in those fields, too, predicted the editor, with proper application and alteration in equipment.[47] Howell had built an anthracite steam engine in which the flames of the coals were not in direct contact with the boiler. Instead, an artificial blast worked by an exhaust bellows increased the heat of the coal burning in a separate firebox and drove the ignited gases into flues under the boiler containing the water. This arrangement eliminated the large firebox necessary for wood or soft coal. At the same time it prevented the flames from striking the boiler plates and subjecting them to the damaging effects of intense heat.[48] The invention became the model for most boilers in the following decade, but was criticized later due to heat loss in the flues.[49]

John Price Wetherill was responsible for the other significant innovation in anthracite stationary steam engines. Wetherill forced a jet of live steam into the firebox. When the steam struck the red-hot

---

[45] *Ibid.*, VI (July-December, 1830), "List of Premiums," p. 8.
[46] James Renwick, *Treatise on the Steam Engine* (New York, 1830), pp. 52-65.
[47] *Journal of the Franklin Institute*, V (1829), 136.
[48] *Ibid.*, pp. 134-135.
[49] W. R. Johnson, *Notes on the Use of Anthracite in the Manufacture of Iron With Some Remarks on Its Evaporative Power* (Boston, 1841), p. 89.

coals it decomposed. The released hydrogen ignited, intensifying the heat and increasing the steam-raising ratio of anthracite fuel.[50]

These inventions were not the only ones to be put on the market. The taproot of the industrial revolution continued to be technological advancement. Improvements on the anthracite stationary steam engine span the middle period.

Coupled with the development of better apparatus was the problem of the comparative efficiency of anthracite and bituminous. This was constantly before the mining interests and the manufacturer. Emulating British fuel experiments with Welsh anthracite, tests were made by American engineers. Dr. Samuel L. Dana, of Lowell, Massachusetts, conducted field experiments with Schuylkill, Beaver Meadow, Lehigh, and Lackawanna anthracite, and compared his findings with similar tests made with Sidney bituminous. These coals were burned under cylinder boilers and the results carefully tabulated. Dana decided in favor of the anthracite as supplying a much better evaporating power than soft coal.[51] The "battle of the fuels" was to be waged for many years, however, and engineers continued to conduct tests in the twentieth century. In 1922 the Bureau of Mines carried out experiments with coke, bituminous, and anthracite in a low-pressure boiler. The Chamber of Commerce of Pittsburgh was vitally interested in the results. The tests concluded that coke and anthracite had greater steam-raising properties in a low-pressure boiler, at a low rate of steaming, but Pittsburgh bituminous proved better for high rates of steaming, which permitted complete combustion of the volatile gases.[52]

Experiments such as Dana's encouraged the anthracite interests. Between 1830 and 1860 large amounts of money were spent explaining the advantages of hard coal in steam engines. The Delaware and Hudson Canal Company, mining the Lackawanna coal, employed an agent at $200 per month, a fabulous sum for the period, to introduce its fuel into New York and New England factories and all establishments where steam engines were employed.[53] The number of stationary steam engines burning anthracite grew slowly, but with a persistence which did credit both to the fuel and the marketing methods. The

[50] *American Journal of Geology and Natural Science.* Reprinted in the *Miners' Journal,* August 20, 1831.

[51] Johnson, *Notes on the Use of Anthracite,* pp. 114-120, 129-136, 156.

[52] John Blizard, James Neil, and F. C. Houghton, *Value of Coke, Anthracite and Bituminous Coal for Generating Steam in a Low-Pressure Cast Iron Boiler,* ("Technical Paper 303, Department of the Interior, Bureau of Mines" [Washington, 1922]), pp. 2-3.

[53] Delaware and Hudson Minute Books, June 20, 1831.

growing popularity of anthracite for steam purposes was reflected even in a negative manner when Philadelphia manufacturers complained about the high prices of coal and threatened to revert to the pine wood from the barrens of southern New Jersey or to purchase Nova Scotia or Virginia bituminous.[54] In the succeeding decades Philadelphia industry relied more and more on anthracite fuel for steam and heat. The city became one of the key centers of eastern manufacturing. Once again location played a significant part in industrial and commercial development. River, rail, and canal routes converged at Philadelphia, carrying raw materials to her mills and factories and finished products from the city to inland markets. More than this, Philadelphia was a thriving port. Overshadowed by New York and challenged by Baltimore, she still was able to maintain a position of eminence along the eastern coast. In the Pennsylvania coal trade alone, Philadelphia exceeded the foreign tonnage of all the seaboard cities, including the port of New York.

The steady supply of Pennsylvania anthracite contributed its share to Philadelphia's industrial development. The steam engine became a leading article of manufacture in nearly all the machine shops of the city. In the late fifties more than a dozen establishments turned out stationary and some portable engines for anthracite coal. None of these shops, however, made steam engines exclusively, nor did they anticipate so large a demand that they kept stocks on hand.[55] The engines were built to order and according to contract specifications. This seems to indicate that the needs of the diversified industry of Philadelphia required a variety of engines which made standardization impractical. Then too, technology was moving so rapidly that new improvements would defeat the value of a backlog of outmoded, expensive equipment. Furthermore, it should be recalled that demand moved with the progress and growth of the industrial revolution which, before 1860 in the United States, had not yet surpassed the agrarian economy.

It is an exaggeration that Philadelphia's steam engines "plied their iron arms in every street," but as their numbers increased in the forties and fifties they burned thousands of tons of coal annually. Ironworks such as Morris, Tasker and Company used extensive amounts of coal for that period. The firm's five steam engines consumed six thousand tons of anthracite each year. Morris, Tasker and Company was one

---

[54] *National Gazette.* Reprinted in the *Miners' Journal,* January 19, 1833.

[55] E. T. Freedley, *Philadelphia and Its Manufactures* (Philadelphia, 1859), p. 316.

of the early coal grate and stove manufacturers in Philadelphia. Its organization grew with the coal trade and the times, producing not only stoves but cast-iron gas and water mains, boiler flues, and gas- and steam-fitter tools. The company continued to manufacture heating equipment, perfecting a self-regulating anthracite hot-water furnace and introducing anthracite steam heat for public and private buildings.[56]

Five large steam sugar refineries, a whalebone factory, textile machinery plants, shipyards, locomotive works, six steam marble mills, steam hat factories, and other industries including glassworks, breweries, brass foundries, boot, shoe, and clothing factories, distilleries, and flour mills used machinery driven by steam produced by Pennsylvania coal.[57]

This list could be applied in whole or in part to New York, Boston, Providence, and to other cities and towns where Pennsylvania anthracite gained prominence. New England imported cotton and coal to her mills.[58] Pennsylvania anthracite passed the one hundred thousand-ton mark in the total imports to Boston in 1841.[59] During the succeeding years anthracite imports increased in Boston, reaching in the fifties an average four hundred thousand tons annually, four times the tonnage of English, Scottish, and Nova Scotia coals.[60] In the same decade the great anthracite mart at New York received nearly six million tons a year.[61] Just how much was used in steam engines and manufacturing processes is difficult to say. Pennsylvania anthracite, always in competition with English, Nova Scotia, and Virginia coals, lost little ground but some prestige to Cumberland bituminous from the Maryland mines. The Cumberland coal, considered to be superior steam fuel by some eastern manufacturers, had entered the market in 1842 when the area was opened by the Baltimore and Ohio Railroad. By 1858 its production totaled a scant six hundred and forty-nine thousand tons, one-third of which was fed into the New York market.[62] Other competitors began to invade the

---

[56] *Ibid.*, p. 322.

[57] *Ibid.*, pp. 141-504.

[58] *Miners' Journal*, April 1, 1848.

[59] Hazard, *U.S. Register*, VI, January 19, 1842, p. 41.

[60] *Boston Daily Advertiser*. Reprinted in the *Twenty-sixth Annual Report of the Philadelphia Board of Trade* (1859), p. 151.

[61] *Annual Report of the Chamber of Commerce for the State of New York. For the Year 1858*, p. 83.

[62] *Ibid.*, pp. 83-85.

East in the fifties; these were Pennsylvania bituminous and semibituminous coals. For the most part, anthracite producers did not consider the domestic competition to be serious. This was the "golden
age" of anthracite in the East. King of the coals, Pennsylvania anthracite scorned the domestic newcomers. There were a few words of
caution offered at the beginning of the decade that the anthracite
empire was being challenged by American soft coals.[63] But a glance
at the comparative production figures, freight rates, and transportation routes assured the anthracite interests that this was sheer
nonsense.

The old Karthaus blacksmith and steam coal, which had moved
down the Susquehanna to tidewater for many years, was bolstered in
1856 by Westmoreland gas coal sent over the Pennsylvania Railroad.
Destined for gasworks, Westmoreland also was used to produce steam
in a few factories. The year before, the semibituminous mines at
Broad Top near Bedford, Pennsylvania, began production. The field
yielded an excellent steam coal. The mines and railroad were developed by Philadelphia capitalists and the area "boomed."[64] But the
amount mined in the five years before the Civil War was small compared with the production of the anthracite districts.

With the large supplies of fuel available to the eastern seaboard,
the great growth in population, and the web of rivers, canals, railroads, and sea lanes connecting the coal fields to the great cities and
smaller industrial towns, it is not surprising to observe that steam produced by coal began serious competition with water power. There
had been little question about steam's superiority over water power
in portions of the West where bituminous was cheap and plentiful.
For years Pittsburgh had used this argument to lure eastern capital.[65]

In New England the factories began to supplement water power
with steam. Even the famous mills at Lowell experimented with
steam power, and also burned Pennsylvania anthracite and English
bituminous under bleaching vats and in their dye houses.[66] Mills at
Providence, Fall River, and Boston spun cloth, rolled iron, and re-

---

[63] D. A. Neal, *Reports Made to the Managers of the Philadelphia and Reading
Railroad Company, 1849-1850.*

[64] *Bedford Inquirer and Chronicle*, November 23, 25, 1855.

[65] *Pittsburgh, Her Advantageous Position and Great Resources, as a Manufacturing and Commercial City* (Pittsburgh, 1845); see also issues of the *Pittsburgh
Gazette*, 1855 to 1860.

[66] *Miners' Journal*, April 3, October 2, 1841.

fined sugar with anthracite steam power.[67] Coal, yielding illuminating gas, had depressed the whaling interests of Salem and New Bedford. Now coal through steam was competing with the millraces of industrial New England. The mineral fuel from eastern Pennsylvania showed increased sales in that area. Steam was in many instances as cheap as water power, and a steam engine was not at the mercy of freezing weather or swollen streams.

In 1845 New York became the target of a concentrated drive to stimulate anthracite steam factories. Redwood Fisher's *National Magazine and Industrial Record* led the movement. Cheap anthracite coal from Pennsylvania and geographical location could combine to make New York the foremost manufacturing city in the United States. Near to the source of industrial steam power, it was also the hub of eastern population. In command of the Hudson Valley and the West through the Erie Canal and the Lakes, the city looked seaward over the coastwise shipping between Boston and Charleston. New York, through coal, was destined to become the greatest metropolis in the world.[68] The enthusiasm did credit both to Redwood Fisher's journalism and to his faith in the continued progress of an American city. Optimism and belief in the future were symptoms of the "Coal Age," for had not the United States boundless resources of this black wealth which had made Britain the giant of industry and power? Unlike Britain, the United States had barely scratched her known coal reserves. The potential strength and future prospects were encouraging to even the most pessimistic.

Stationary steam engines and ironworks consumed hundreds of thousands of tons of anthracite, but the coal was used in a number of other ways, too. William Everhart employed anthracite in his brick kiln in 1831 and is credited with being the first to use the fuel successfully in that enterprise.[69] Anthracite found utility in agriculture when it was used to burn lime. Limestone yielded about one hundred bushels to every ton and a half of coal.[70] In the twenties and thirties the coal and the stone were burned together. The result was a rather poor lime filled with ash, clinker, and other impurities. It was not until 1840 that a method was discovered that would produce pure

---

[67] Hazard, *Register of Pennsylvania*, XVI, 348; *The North American Review*, XLII, 255; *Fisher's National Magazine and Industrial Record*, III, June, 1846, pp. 46-47.

[68] *Fisher's National Magazine*, I, September, 1845, pp. 358-362.

[69] Hazard, *Register of Pennsylvania*, VIII, 192; *Miners' Journal*, September 17, 1831.

[70] Hazard, *Register of Pennsylvania*, I, 312.

lime with anthracite as fuel. The coal was placed on a grate and a blast of air was applied by a fan worked by a small steam engine. The blaze was then blown throughout the lime kiln, the coal ash falling into the pit below the grate. A kiln could be burned through in thirty hours, while with wood it took fifty-four.[71] The use of anthracite coal in the lime kilns of the Lehigh Valley not only aided agriculture but formed a workable basis for the thriving cement industry of later years.

Novel, if somewhat impractical uses for anthracite often appeared in print: ashes could be used for fertilizer or for cleaning brass, and pulverized coal would make fine shoe blackening or worm repellent for fruit trees. Coal region contractors burned anthracite in seasoning green lumber, and it was used in a similar way to cure tobacco. From coloring hats to carving bric-a-brac, anthracite supplied the need. In the latter case, a one hundred thousand-dollar novelty firm chartered by the State operated under special patent.[72]

In 1840 *The Pennsylvanian* called attention to a full-sized speaking trumpet for the Philadelphia Fire Department which was carved from a block of anthracite coal. The newspaper commented that "in Pennsylvania, no embellishments can be more appropriate than those made from one of the great staples of the state."[73] It was a far call from a carved trumpet to the genius and perseverance which brought anthracite to maturity in the factories of the East. The symbolism of the trumpet persisted as a part of the age, and through it passed the sounds of industry powered by steam and fed with Pennsylvania coal.

The stationary steam engine burning anthracite coal was found not only in the factories of Philadelphia, New York, Boston, and Lowell, but it spread to the smaller industrial towns along the transportation routes of river, canal, and railroad.[74] It also was found to be indispensable in the anthracite coal regions. Here the manufacture and use of anthracite steam machinery was of particular importance as the equipment was employed in the very industry from which it drew

---

[71] Hazard, *U.S. Register*, II, June 17, 1840, p. 399.
Carolina Waters, 1826-1836. Porthmouth, Virginia, 1849.

[72] Varied uses for anthracite appeared in the literature of the period. Hazard, *Register of Pennsylvania*, VII, 192, IX (January-June, 1832), 160, XV (January-June, 1835), 308; *The North American Review*, XLII, 255. Hunt's *Merchants Magazine*, VI (January-June, 1842), 194; *Journal of the Franklin Institute*, XII, 5-6; J. L. Bishop, *A History of American Manufactures* (Philadelphia, 1864), II, 362.

[73] *The Pennsylvanian*, February 21, 1840.

[74] *Miners' Journal*, October 23, 1841.

its energy. The North American Coal Company was the first mining organization in Schuylkill County to use steam-powered equipment. A fifteen-horsepower steam engine, built in 1833 by the firm of Haywood and Snyder, was employed in pumping water and hoisting coal. The pumps ran fifteen hours during the twenty-four with a capacity of 440 gallons per minute at one hundred feet. For the remaining nine hours, a few hours short of the miner's working day, the steam power was shifed to the hoist and the coal was drawn from the mine. The engineer, an Englishman with many years of experience with bituminous in England, preferred the anthracite, which he claimed could generate steam more regularly than any other coal. The fuel consumption of the North American company's engine was about two tons of coal every twenty-four hours.[75]

The successful application of anthracite fuel to steam-powered mining machinery was not of immediate importance to the coal regions. It was not until the mid-forties that some anthracite mines were confronted with serious problems of flooding.[76] Slopes and drifts were cut into hillsides and followed the coal seams on a horizontal as far as possible. Later, when shaft mining was employed on a large scale, plunging verticals deep into the earth and sending radial tunnels into the mountain, underground springs and streams menaced the works. At the same time, coal was hoisted to the surface of a shaft mine with difficulty. Animal power was insufficient, but steam provided the solution. These changing methods in mining promoted the increased production and use of steam-powered equipment in the form of hoists and pumps.

In 1839 Schuylkill County, the heaviest coal production region in the anthracite fields, listed only thirteen steam engines employed in mining operations.[77] At that time steam equipment was still something of a novelty in the coal regions, and when a new engine was placed in operation at a mine it was usually attended by a celebration. Pottsville launched her steam engines with the ceremony due a vessel of the deep. The press, city officials, and other local dignitaries usually attended the affair when a new steam pump and hoist was introduced. At the Gate Vein near Pottsville in 1841, speeches

[75] *Ibid.*, August 21, 1834; Hazard, *Register of Pennsylvania*, XIV (July-December, 1834), 144.

[76] *Second Annual Report on the Geological Exploration of the State of Pennsylvania* (1838), p. 87; *Eighth Census, Manufactures*, p. clxxiii.

[77] Hazard, *U.S. Register*, I, November 20, 1839, p. 252.

were made eulogizing the present importance and future prospects of anthracite coal. Refreshments were consumed in quantity while the success of the engine was toasted by all in great gulps of Yuengling's best ale.[78]

The forties saw the gradual application of steam pumps, hoists, and breakers to the anthracite industry. A new era was opened in Schuylkill County. Machine shops in Pottsville, Minersville, and adjoining towns began turning out steam equipment which, for the most part, was sold within the coal regions.[79] As the coal traffic grew, a pressing problem was to find an efficient and rapid method for breaking coal into various sizes for market. A breaker invented by J. S. Hubbell in 1836 met with a cool reception and was called impractical by the Franklin Institute.[80] There was little success in perfecting this invention until 1844 when the Battens erected a steam breaker in their retail yards at Philadelphia. Not long after this, Gideon Bast, of Minersville, operating a mine on Wolf Creek, constructed a steam breaker and screening device which did the work of forty to fifty men, breaking and screening two hundred tons of coal a day.[81] Prior to these inventions coal had been broken by hand. The anthracite steam breakers could perform the task for eight to ten cents per ton, while the cost of hand labor was estimated to be thirty to thirty-seven cents per ton.[82] More than this, the perfection of the breaker encouraged the greater use of anthracite fuel in stationary steam engines as the small pea coal, preferred by steam engineers, was now available in larger quantities at cheaper rates. In 1848 the Schuylkill region had eighty-four breakers in operation, and of this number seventy-three were steam powered.[83] In that same year Schuylkill Valley mine owners employed 167 steam engines in their industry.[84] Five years later the number had grown to 205, and this did not include thirty-three additional steam engines used for other purposes. Sawmills, gristmills, rolling mills, machine shops, and even the new press

---

[78] *Miners' Journal,* October 23, 1841.

[79] HSSC, Parry Papers, Memoranda of agreements for the construction of steam engines: Benjamin F. Pomeroy and the Delaware Coal Company, July 15, 1841; J. D. Steinberger and J. Sillyman and Company, February 17, 1848; William De Haven and David D. Lewis (for partnership in steam engine factory), February 8, 1848. See also Hazard, *U.S. Register,* V, September 15, 1841, p. 117.

[80] *Journal of the Franklin Institute,* XXI (January-June, 1836), 202.

[81] HSSC, Baird Halberstadt Papers, W. H. Egle, M.D.

[82] *Miners' Journal,* May 25, 1844.

[83] Eli Bowen (ed.), *The Coal Regions of Pennsylvania* (Pottsville, 1848), p. 38.

[84] *Ibid.*

which printed Benjamin Bannan's *Miners' Journal* were tied to the strength of steam and anthracite coal.[85]

By 1856 Schuylkill County claimed more steam engines than any area in the State with the exception of Philadelphia.[86] Steam-powered mining equipment and steam engines for manufacturing multiplied in the adjoining anthracite fields of the Wyoming, Lehigh, and Lackawanna valleys, contributing to the increased production of coal through steam and adding to the consumption of the fuel mined to supply their boilers.

---

[85] *Miners' Journal,* January 8, 1853.
[86] *Ibid.,* April 26, 1856.

CHAPTER IV

Fuel Problems
of the Iron Industry

The Challenge of Anthracite

Today toast making is a lost art, but a century ago wherever men gathered around a banquet table, glasses were sure to be drained many times during the course of an evening's festivities as the toasts were proposed.

At a celebration in Pottsville over a hundred years ago, Nicholas Biddle, after speeches had been duly made, arose to his feet, lifted his glass high and drank to "Old Pennsylvania—Her sons, like her soil—a rough outside, but solid stuff within;—plenty of Coal to warm her friends—plenty of Iron to cool her enemies."[1] The occasion was in honor of the successful smelting of iron ore with Pennsylvania anthracite coal, an event to be of great significance to the eastern Pennsylvania economy over the following twenty-five years.

Unsuccessful attempts to smelt iron ore with anthracite coal had been made twenty years prior to the Pottsville triumph. In 1820 the Lehigh Coal and Navigation Company had experimented with anthracite in its blast furnaces at Mauch Chunk. At the same time similar experiments were carried on by two Frenchmen, Gueymand and Robin, at Vizille, France, a small town near the Swiss border. In both instances a cold blast was used on the charge of coal, ore, and limestone. The tests failed as a result of the incomplete combustion of the fuel.[2] Despite the discouraging results, the idea of using anthracite coal as a fuel for the blast furnace was not discarded. To Americans who accepted the challenge of anthracite and believed in its utility, it seemed inconceivable that a fuel containing over ninety per cent carbon could not be adapted to the blast furnace.

Joshua Malin, one of the leading advocates of anthracite in the iron industry, designed an anthracite furnace for smelting ore and sent the

---

[1] Bowen, *The Coal Regions of Pennsylvania*, p. 33.
[2] Johnson, *Notes on the Use of Anthracite*, pp. 14-26.

61

description to the Franklin Institute in 1827, but he could not convince the Institute nor the ironmasters of its practicality.[3] Not long after the opening of the Schuylkill coal region, the Pottsville *Miners' Journal*, seeking to expand the coal trade through new applications, printed an article from the *American Journal of Science and Arts* written by Dr. William Meade, who related his experiences with Rhode Island anthracite. Meade had bribed workmen to use anthracite at a Kingston, Massachusetts, ironworks which smelted bog ore. Under his directions the furnace was charged with alternate layers of ore, charcoal, and Rhode Island anthracite. According to the article the ore was smelted successfully. Meade felt that Pennsylvania ironmasters had not given anthracite a fair trial and urged them to abandon their prejudices and experiment with it in their furnaces. Both the *Miners' Journal* and Dr. Meade suggested that the anthracite be broken into small fragments to insure more complete combustion.[4]

Meanwhile British ironmasters sought improved methods in smelting iron ore with coke and raw bituminous coal. Obtaining a patent in 1828, James B. Neilson, superintendent of the Glasgow Gas Works, successfully employed the principle of the hot-air blast. Air was heated to 300° F. and blown into the furnace. After patenting the improvement in both England and France, he carried on further experiments at the Clyde Iron Works. The hot-air blast was one of the most significant discoveries in the iron industry, resulting in larger yields of pig iron with less fuel by causing more complete combustion of the latter and a union of a greater proportion of carbon to the metal. The hot blast first was applied to coke, but experiments showed that it could be employed also to smelt ore with raw bituminous coal.[5] This was demonstrated by Neilson and William Dixon at the Calder Iron Works in 1831. Dixon raised the temperature of the blast to 600° F., employed water-cooled tuyères or nozzles to deliver the blast, and produced one ton of iron using less than three tons of bituminous coal. The hot blast was a remarkable improvement over the cold blast which, to smelt the same amount of iron, needed coke produced from eight tons of bituminous coal.[6] Neilson estimated the annual saving in fuel costs to the British iron industry

---

[3] *Journal of the Franklin Institute*, IV (1827), pp. 217-219.

[4] *Miners' Journal*, April 14, 1827.

[5] J. B. Neilson, "On the Hot Air Blast," *Transactions of the Institution of Civil Engineers*, I (1836), 23-24.

[6] *Journal of the Franklin Institute*, XXIV (1837), pp. 46-52.

at £296,000.[7] Within ten years the hot-air blast dominated English ironmaking. Applied principally to coke, the chief fuel of the English blast furnace, it was used also whenever raw bituminous coal, or later, anthracite, was employed to smelt the ore.[8]

Interest in smelting iron ore with mineral coal gradually made headway in Pennsylvania during the decade of the thirties. The Franklin Institute in 1831 and 1832 offered medals to anyone successfully smelting ore and producing at least twenty tons of iron using raw bituminous or anthracite coal,[9] and Hazard's *Register of Pennsylvania* in 1835 suggested that the legislature award premiums for the smelting of iron with anthracite coal.[10] During the following year Charles B. Penrose, of Cumberland, Perry County, Pennsylvania, sponsored a bill, which became law on June 16, 1836, to encourage the manufacture of iron from coke or mineral coal by authorizing the governor to form a joint stock company for that purpose.[11] In 1835 Pennsylvania provided for the formation of a geological survey for the specific objective of learning more about the coal and iron deposits of the State.[12] Belief in the future success of the experiments with mineral fuel being carried on by some Pennsylvania ironmasters and in the triumph of "American ingenuity" was expressed by Governor Ritner in his annual message to the Pennsylvania General Assembly in 1837.[13] The next year Ritner, a little prematurely, announced to the assembly that experiments to smelt iron ore with anthracite coal at Mauch Chunk, Manayunk, and Easton had proved

> So perfectly satisfactory . . . that large furnaces, in which anthracite coal alone is to be used as fuel, are now in progress of construction, at several points in the state. The successful union of stonecoal and iron ore, in the arts, is an event of decidedly greater moment to the prosperity of our state, than any that has occurred since the application of steam in aid of human labor.[14]

---

[7] Neilson, "On the Hot Air Blast," *Transactions of the Institution of Civil Engineers,* I, 23-24.

[8] *Documents Relating to the Manufacture of Iron in Pennsylvania* (Philadelphia, 1850), p. 33.

[9] A. C. Bining, "The Rise of Iron Manufacture in Western Pennsylvania," *The Western Pennsylvania Historical Magazine,* XVI, 246; *Journal of the Franklin Institute,* XII, 1, 233.

[10] Hazard, *Register of Pennsylvania,* XIV, 414-415.

[11] *Pennsylvania Senate Journal,* 1835-36, I, 661, 1057-58.

[12] *Pennsylvania Archives,* Fourth Series, VI, 473.

[13] *Ibid.,* p. 391.

[14] *Ibid.,* p. 474.

Despite the oratorical optimism of the Pennsylvania Governor, "the successful union of stonecoal and iron ore" was not yet as successful in the State as he had led the assembly to believe. Complete triumph was not to be realized until the winter of 1839. On the other hand, Ritner had good reason to sound encouraging. He not only could point to the use of mineral coal in British furnaces as proof of adaptation, but to the growing progress in Pennsylvania in the experiments with anthracite iron. British trials with raw bituminous and the hot blast were striking examples of the profitable use of soft coal in the blast furnace. In 1837 George Crane in Wales had applied the principle of the hot blast to Welsh anthracite with acknowledged success. In Pennsylvania empirical tests with pure anthracite coal or anthracite mixed with wood, charcoal, or bituminous had been carried on for several years. Considerable headway had been made with pure anthracite by a Protestant clergyman, Dr. Frederick W. Geissenhainer, of New York City, who had applied the hot blast to Schuylkill coal in 1833 at the Valley Furnace near Pottsville. His findings were published in the *Journal of the Franklin Institute,* but the editor seemed suspicious of the doctor's claims: "The patentee states that he has discovered the true principle heretofore unknown, of applying anthracite to the purpose of smelting, etc.; and we sincerely hope he may find his theory or rather his practice correct."[15] Geissenhainer's experiments at the Valley Furnace showed promise, but were not completely successful.[16] He later sold his patent to George Crane who, as previously noted, had successfully applied the hot blast to anthracite in Wales in 1837.[17] Crane obtained patents in Great Britain and the United States, but soon became entangled in a series of lawsuits concerning his process.

In Pennsylvania, meanwhile, determined ironmasters felt their way slowly toward the solution of using anthracite coal in the blast furnace. William Lyman, of Boston, was made superintendent of the experimental anthracite Pioneer Furnace at Pottsville in 1839. Disappointment and discouragement over the failure to produce anthracite iron that summer caused the proprietors to employ Benjamin Perry. The new man assumed the responsibility and brought in a successful blast in October of the same year. A former furnace manager of the

---

[15] *Journal of the Franklin Institute,* XVII, 395-396.

[16] J. W. Swank, *Introduction to the History of Iron Making and Coal Mining in Pennsylvania* (Philadelphia, 1878), pp. 73-74.

[17] George Crane, "On the Smelting of Iron With Anthracite Coal," *Journal of the Franklin Institute,* XXV (1838), 126-129.

Pentyweyn Iron Works in Monmouthshire, South Wales, Perry had come to the United States in 1838 to conduct experiments with bituminous coal and coke at Farrandsville, Pennsylvania. He was familiar with both Neilson's and Crane's hot-blast processes.[18] The hot blast of the Pioneer Furnace was blown by steam power and heated to 600° F. by anthracite coal in ovens at the base of the furnace. The furnace was kept in continuous blast for over three months, producing about twenty-eight tons of anthracite pig iron each week. Recognized success had been attained at last! It was to commemorate this achievement that Nicholas Biddle, in January, 1840, made an award of $5,000 to the Pioneer Furnace erected by Burd Patterson and supervised first by Lyman and then by Perry.[19] Biddle's speech and toast to the occasion, along with the cash award, were given wide publicity in the press and remain indicative of the importance the business interests of eastern Pennsylvania placed upon the successful smelting of iron ore with anthracite coal.

Since the beginning of the decade of the thirties, eastern Pennsylvania business interests had viewed with misgivings the rising costs of charcoal for smelting iron. The average price of charcoal in 1839 was five cents per bushel. It was estimated that it took two hundred bushels at a cost of ten dollars to produce one ton of pig iron. Anthracite coal could be purchased for about $2.50 a ton and for less than that near its source of supply. A little more than two tons of anthracite coal were consumed in the blast furnace to produce one ton of anthracite pig iron. Fuel costs were cut nearly fifty per cent.[20] This reduction was not immediately realized, but the cost of coal to the ironmaster, who purchased it in large quantity, did not vary greatly to 1860, holding an average price level over the years of three dollars per ton.[21]

With the reduced cost of production, the iron interests felt they were in a better position to meet the increased competition of British

[18] Hazard, *U.S. Register*, I, July 3, 1839, p. 27.

[19] Swank, *Iron Making and Coal Mining in Pennsylvania*, pp. 73-75; Bowen, *The Coal Regions of Pennsylvania*, p. 33; *Niles' National Register*, LVII (January-June, 1840), 57, 176, 313, 386; Hazard, *U.S. Register*, II, April 8, 1840, pp. 230-231; *The Pennsylvanian*, January 29, 1840.

[20] *Poulson's American Daily Advertiser*, January 11, 1839; Hazard, *U.S. Register*, I, November 20, 1839, p. 352; HSP, C. G. Childs, Coal Used in Making Iron, Twenty-four Notebooks, 1845-1865.

[21] Chester County Historical Society, Pennypacker Papers, Balance Sheet, January 1, 1853, The Working Men's Nail and Iron Company, Phoenixville, referred to hereafter as CCHS. C. M. Wetherill, *Report on the Iron and Coal of Pennsylvania* (1850).

iron imports. Those who supported the tenets of economic nationalism applauded the new use of anthracite as a second Declaration of Independence hurled at Britain's monopoly of the iron trade. Nicholas Biddle, in his famous speech at the Mount Carbon Hotel in Pottsville, pointed to the flight of American money in the purchase of iron from Great Britain in 1836-37. "This dependence is deplorable," said Biddle. "It ought to cease forever; and let us hope that with the new power, this day acquired, we shall rescue ourselves from such a costly humiliation."[22]

The use of anthracite coal in the smelting of iron ore had another effect on the economic scene. In order to be near a plentiful supply of forest wood for charcoal used in the blast furnaces, iron plantations were located some distances from population centers and markets and often miles from the major transportation routes. The anthracite coal trade, by 1840, had already developed improved river navigation and canals from the coal fields to tidewater. Railroads soon followed the river valleys, paralleled the canal lanes, or cut iron paths of their own through the anthracite regions. Anthracite furnaces grew up along these established coal trade routes close to or within the limits of the towns. Danville, Scranton, Trenton, Allentown, Harrisburg, Columbia, and other urban areas erected furnaces which no longer depended upon vast acres of woodland, but received their fuel from the mines of eastern Pennsylvania.[23] The Lehigh Valley became the center of the anthracite iron industry. But in 1845 the Pottsville *Miners' Journal,* carried away by the rapid growth of anthracite furnaces along the Schuylkill, visualized in twenty years a continuous manufacturing town between Pottsville and Philadelphia founded on anthracite steam engines and anthracite iron.[24] Transportation routes aided in supplying the furnaces with the fuel, limestone, and iron ore, all products of eastern Pennsylvania. Over half the iron ore, however, was supplied to the furnaces by farmers who dug it on their own property during the winter and hauled it to the furnaces in their locale.[25]

The successful achievement of Pennsylvania ironmasters in smelting iron with the hot blast and anthracite coal caused a rapid growth in the numbers of anthracite furnaces east of the Alleghenies. A short

---

[22] *Merchants' Magazine,* XVI, 588.

[23] *Ibid.,* p. 592.

[24] *Documents Relating to the Manufacture of Iron,* pp. 106-107.

[25] Reprinted in the *American Railroad Journal,* XVIII (1845), p. 787.

time before the Pottsville furnace experienced good fortune, Baughman, Guiteau and Company, using the Lehigh Coal and Navigation Company's old furnace at Mauch Chunk, had obtained sporadic yields of anthracite iron.[26] The Biddle-Chambers furnaces at Danville, Reeves's at Phoenixville, Burd Patterson's at Roaring Creek, George Patterson's at Danville, and others followed Pottsville's example in quick succession during the spring and summer of 1840. The next year Pennsylvania counted a dozen anthracite furnaces and New Jersey had four at Stanhope along the Morris Canal.[27] The Stanhope furnaces led the *Newark Daily Advertiser* to boast that New Jersey would become Pennsylvania's competitor for at least one-half the world's iron markets.[28] By 1846 there were forty-three anthracite blast furnaces in Pennsylvania, using large amounts of coal and producing approximately one-third of the State's pig iron. In ten years the number had doubled.[29]

The Lehigh Coal and Navigation Company, founded by Josiah White and Erskine Hazard, had been one of the pioneers in the attempts to make anthracite iron. Discouraged over the repeated failures through the years and ignored when they offered to provide free toll, water power, and cheap coal to anyone successfully smelting ore with anthracite in the Lehigh region, the board of managers made plans to create a subsidiary company operating under the Crane patent.[30] Erskine Hazard and the young Solomon Roberts, a nephew of Josiah White, sailed for Wales in 1838 and attempted to persuade George Crane to come to the Lehigh Valley and set up an anthracite furnace. Crane, deep in lawsuits over his patents, could not make the trip and it was suggested that one of his foremen, David Thomas, go in his place. A contract was drawn up by Hazard. Thomas agreed to its terms which permitted him two years to build a successful blast furnace for smelting iron ore with anthracite.[31] Under the name of

---

[26] Swank, *Coal Mining and Iron Making in Pennsylvania,* p. 75; Johnson, *Notes on the Use of Anthracite,* pp. 33-35; *Poulson's American Daily Advertiser,* January 11, 1839.

[27] Johnson, *Notes on the Use of Anthracite,* p. 27.

[28] Reprinted in Hazard, *U.S. Register,* V, October 6, 1841, p. 216.

[29] Bining, "The Rise of Iron Manufacture in Western Pennsylvania," 247.

[30] *Report of the Board of Managers of the Lehigh Coal and Navigation Company to the Stockholders,* January 3, 1840. Reprinted in Hazard, *U.S. Register,* II, March 4, 1840, p. 157.

[31] *Ibid.*; Eleanor Morton, *Josiah White, Prince of Pioneers* (New York, 1946), p. 214.

the Lehigh Crane Iron Company, whose stockholders included White, Hazard, the Earp family, Timothy Abbot, John McAllister, and Nathan Trotter, the furnace was erected by Thomas at Catasauqua, near Allentown. It was put into blast on July 4, 1840, and soon produced fifty to sixty tons of foundry iron a week.[32] The Lehigh Crane Iron Company became the leading anthracite iron establishment in Pennsylvania, producing approximately one-tenth of the anthracite pig in the State. The following figures indicate the validity of this claim:[33]

Lehigh Crane Iron Company*

Year	Tons of Anthracite Consumed	Tons of Anthracite Pig Iron Produced
1856	67,900	31,094
1857	66,500	30,943
1858	60,800	28,870

Tons of Anthracite Pig Iron Produced	1856
Pennsylvania	306,972
All others U. S.	87,537

* Reported by J. P. Lesley, Secretary of the American Iron Association.

The supremacy of the Lehigh Crane Iron Company was challenged by the extensive Montour Iron Works at Danville, using Wyoming Valley anthracite. By 1857 Montour's four blast furnaces were making twenty-four thousand tons of anthracite pig iron a year. The Montour works included two rolling mills producing over twenty thousand tons of finished iron rails a year. The blast furnace, rolling mills, and their puddling furnaces consumed annually an estimated one hundred thousand tons of anthracite coal.[34]

In the forties Lancaster County had a small boom in anthracite iron. The fever was contagious and it seemed that almost everyone wanted to be an ironmaster. Even before the successful use of anthracite in the iron industry, Lancaster County had petitioned Congress for the establishment of a national foundry. The foundry, to be located on the Conestoga Creek, would make arms, cannon, and castings to supply the military establishments of the United States. The inland location would be of major strategic importance. With advancements in naval warfare, it was argued that coastal towns no

---

[32] Hazard, *U.S. Register,* II, March 4, 1840, p. 157; III, August 26, 1840, p. 142.
[33] *Twenty-sixth Annual Report of the Philadelphia Board of Trade* (1859), p. 115.
[34] *Pittsburgh Gazette,* January 9, 1857.

longer could be defended against bombardment. Here, in the rolling hills of Lancaster County, the government of the federal union could forge the sinews of war without the danger of destruction by the guns of an enemy navy. Local sources of iron ore from the Chestnut Hill, Conowingo, and Cornwall mines, all within the radius of twenty miles of the city of Lancaster; and convenient access to anthracite coal supplies hauled over canal and river routes forty to seventy miles, provided other strong points of argument.[35] Although the House of Representatives ordered five thousand copies of the memorial to be printed, there was no further action by the federal government. Lancaster County, however, began constructing hot-blast anthracite furnaces within four years after the Pottsville experiments had proven successful. The first furnaces in the county to use anthracite were the Shaunee at Columbia in 1844 and the Henry Clay and Chikiswalungo in 1845-46, followed a few years later by the Conestoga, Donegal, Marietta, and others. The most famous of these was the Chikiswalungo. Built originally by Henry Haldeman and put into blast in July, 1846, it was never out of blast for more than six months at a time until 1893.[36]

The fuel burned by the anthracite furnaces was carefully selected for its performance under blast, although preference was sometimes altered by problems of transportation and supply. The theory that red-ash anthracite contained more sulphur than white-ash was exploded by careful fuel analysis. Sulphur was a particular hazard in iron manufacture as it made the iron brittle. So complete was the analysis, testing twenty-three samples of white- and red-ash varieties, that it discovered some of the red-ash from the Pottsville area had even less sulphur than the white.[37] But the difference in the amount of sulphur content was so small that it had no bearing on the ironmasters' selection of anthracite for the blast furnace. The free-burning red-ash from Pine Grove, so popular in home and factory in New York, could be used in blast furnaces, but was not preferred. Instead, ironmasters preferred white-ash because it contained more carbon and burned more slowly than red-ash. A hard, durable coal, it also withstood shipment and handling without crumbling and attendant loss. The Big Vein at Wilkes-Barre mined by the Baltimore, Diamond,

[35] *The Voice of Lancaster County Upon the Subject of a National Foundry* (Lancaster, 1839); F. S. Klein, *Lancaster County, 1841-1941* (Lancaster, 1941), pp. 9-10.

[36] H. L. Haldeman, "The First Furnace Using Coal," *Lancaster County Historical Society Papers*, I (1896), 14-23.

[37] Taylor, *Statistics of Coal*, pp. 416-417.

and Black Diamond coal companies furnished the white-ash anthracite for the Marietta, Donegal, Clay, and Chickiswalungo furnaces.[38] The Montour Iron Works at Danville and former Governor Porter's furnaces at Harrisburg also used anthracite from the Wyoming basin.[39] Furnaces in the Lehigh and Schuylkill valleys used local white-ash, most varieties of which were considered excellent for smelting ore.[40]

Anthracite ironmasters were practical businessmen. They constantly sought to improve their product and increase their profits by adopting better methods in smelting ore. Experiments with the hot blast and antracite fuel continued.[41] The amount of fuel was reduced somewhat by heating the blast with the hot gases which escaped from the furnace. The boilers of the steam engine supplying the force of the blast were heated in the same manner.[42] Numerous tests were made on the comparative strengths of cold-blast and hot-blast iron. Although the merits of both types of iron were argued at length by their supporters, there was general agreement on two points: cold-blast charcoal iron was the better metal, but anthracite iron cost less to make. The economy in the manufacture of the latter rested upon cheap fuel. Hot-blast anthracite gained steadily in production over cold-blast charcoal.[43]

The year 1856 found Pennsylvania-anthracite pig iron production to be more than three times the charcoal pig iron production in the State. Besides this, it had outdistanced by more than fifty thousand tons the charcoal pig iron production of the rest of the United States. The amount of anthracite pig iron for the United States totaled nearly four hundred thousand tons, of which over three hundred thousand had been smelted in eastern Pennsylvania furnaces.[44] These figures do not mean that the charcoal furnace was considered a relic of the past. On the contrary, new charcoal furnaces were built, especially in western Pennsylvania and in other areas where wood was plentiful and supplies of anthracite impossible to obtain. But some charcoal furnaces were constructed in eastern Pennsylvania, the heart

---

[38] Ibid.

[39] Wetherill, Report on the Iron and Coal of Pennsylvania.

[40] Taylor, Statistics of Coal, p. 424.

[41] Journal of the Franklin Institute, XLVII (January-June, 1849), 393-402.

[42] American Journal of Science and Arts, Second Series, VI (November, 1848), 74-80.

[43] Journals of the Franklin Institute, XXIX (January-June, 1840), 62; XXXV (January-June, 1843), 134-36; XXXVII (January-June, 1844), 111, 126; XXXVIII (July-December, 1844), 117-121.

[44] See Tables C and D, Appendix.

of the anthracite iron industry.[45] Demand for the more expensive charcoal pig still was high because it was used in the manufacture of finer types of iron work. These needed a tough, malleable metal of uniform quality to satisfy the agrarian market. Anthracite iron found its major market in iron rails, boiler plate, and heavy machinery, which, with the new industrial expansion, made large demands on the anthracite furnaces.[46] By 1858 there were 120 anthracite furnaces in the United States. Pennsylvania had ninety-two while New York was next with fifteen. There were five of them in New Jersey, four in Maryland, three in Massachusetts, and one in Connecticut.[47]

Anthracite was used not only in the blast furnace, but in the other processes of the iron industry. From the ore to the finished product anthracite was employed by increasing numbers of eastern ironmasters. Many blacksmiths, too, used anthracite coal in their forges. About 1770 two enterprising brothers from Connecticut, Obediah and Daniel Gore, who had settled at Wilkes-Barre, discovered and used anthracite in their blacksmith shop. This usually is noted as the first practical use of "stone coal" in Pennsylvania.[48] From these early experimental beginnings the knowledge of anthracite's utility gradually spread among the blacksmiths of the region.

During the American Revolution gunsmiths at Carlisle employed anthracite mined from the Hollenbach bed, one mile above Wilkes-Barre. This fuel was hauled with great difficulty on water and overland to its destination, evidence of its esteem in the eyes of these patriot artisans.[49] Jesse Fell, of Wilkes-Barre, inventor of a famous grate, also used anthracite coal in his nailery in 1788.[50] Thus, half a century before the development of the anthracite coal trade, the fuel had attained a local reputation for usefulness, overcoming the provincial belief that it would not burn in the smith's forge. This local reputation was gained also in the Lehigh and Schuylkill valleys, where smiths found anthracite to be an economical fuel which could heat iron in half the time of charcoal. One ton of anthracite was equal to

---

[45] *Report of a Committee to the Iron and Coal Association of the State of Pennsylvania* (Philadelphia, 1846), pp. 11-13; *Twenty-Sixth Annual Report of the Philadelphia Board of Trade* (1859), p. 121.

[46] *Pennsylvania Archives,* Fourth Series, VI, 722.

[47] J. P. Lesley, *The Iron Manufacturers Guide to the Furnaces, Forges and Rolling Mills of the United States* (New York, 1859), pp. 1-24.

[48] HSP, Jesse Fell to Johnathan Fell, December 1, 1826.

[49] HSP, W. J. Buck, Early Discovery of Coal, MS.

[50] HSP, Jesse Fell to Johnathan Fell, December 1, 1826.

approximately two hundred bushels of charcoal in the smith's forge. Many smiths preferred it, too, as anthracite was a clean fuel which did not smudge and begrime them as did the charcoal.[51] The use of anthracite as a blacksmith fuel spread to the towns and cities affected by the anthracite coal trade, and though it did not replace charcoal entirely, it found an increasing market in the smith shops of the East.

About 1803 Joshua Malin, to become one of the foremost pioneers in the use of anthracite coal in the iron industry and in steam engineering, was employed at Samuel Mifflin's Philadelphia slitting mill where he used an experimental mixture of anthracite and coke to heat iron. The anthracite had come from Lehighton. Malin's account, written a quarter of a century later, does not mention the source of the coke.[52] There is no evidence that the coke of bituminous coal was made in the United States at this early date, but it would not have been impossible for a recent immigrant from the British Isles, skilled in coke making, to have been employed by Malin for that purpose. It would have been very difficult to import the fuel from England by sailing ship, as coke crumbles easily. There is the possibility, however, that Malin was referring not to the coke of bituminous coal, but to the coke of forest wood, or simply charcoal. Whatever the source or kind of coke, Malin seemed well pleased with the results of his tests. But because of the expense entailed in hauling the anthracite from its isolated valleys, he soon returned to the use of Virginia and English bituminous, which were more economical despite the distances of transport by sea.[53]

During the War of 1812 the blockade provoked a scarcity of Virginia and English coals. Malin, then a partner of Armor Bishop, operated an iron works in Delaware County. He made a contract with Colonel George Shoemaker to haul coal overland by wagon from the Schuylkill area to the mill works. When Shoemaker raised the price of the anthracite, the enterprising Malin opened his own pit in the region near the site of Pottsville and was able to obtain the fuel at cheaper rates. In two years twenty thousand bushels, or about seven hundred tons, were hauled overland by wagon to Delaware County to heat the iron for Malin's slitting mill. At this time, between 1812 and 1814, Malin conducted a number of interesting experiments with anthracite, foreshadowing the growth of Pennsylvania's anthracite

---

[51] *Journal of the Franklin Institute,* X (July-December, 1830), 198-201.
[52] HSP, Joshua Malin to Gerard Ralston, April 20, 1827.
[53] *Ibid.*

iron industry of the forties.[54] Experiments in heating iron were no doubt carried out at the mill, but in 1813 Malin also purchased a half interest in the Valley Forge on the Schuylkill in Chester County where he successfully melted pig iron in an air (remelting) furnace with anthracite coal, made several hundred tons of castings, and claimed the distinction of being the first man in the United States to have accomplished this.[55]

Malin's successful experience with anthracite in his slitting mill encouraged others to use the fuel. Josiah White and Erskine Hazard purchased some for their wire works at the Falls of the Schuylkill. Here, quite by accident, the principle of the minimum draft was discovered when a discouraged workman, who had been trying all day to make the coal burn properly, closed the furnace door in disgust and went home. He returned a few hours later for his forgotten jacket and noticed the fire burning brightly.[56] So impressed were White and Hazard over the use of anthracite in their industry that they foresaw a great future for the fuel if it could be made available in quantity at cheaper prices. It was the successful use of the coal at the Falls of the Schuylkill that gave these two men the idea to form a corporation to improve the navigation of the Schuylkill for the purpose of bringing coal to tidewater. The Pennsylvania legislature scoffed at their idea to bring "black stone" to Philadelphia. When the Schuylkill Navigation Company was finally incorporated in 1815, Josiah White did not have any controlling interest. The tolls on coal had been fixed by charter at thirty cents a bushel or $8.40 per ton. The managers refused to alter the amount. White threatened to use the Lehigh, a hazardous, turbulent river without any slackwater or canal improvements. Within a few years White carried out his threat and formed with Hazard and a third party, George Hauto, the Lehigh Navigation Company. The firm began mining anthracite and transporting it to Philadelphia in 1820.[57]

The use of anthracite in the manufacture of cast iron was essentially an American improvement. This was also true of the use of anthracite in heating iron for the rolling mills and in puddling pig iron.[58] Puddling, or the melting and stirring of the molten pigs to

---

[54] *Ibid.*

[55] *Ibid.*

[56] HSP, W. J. Buck, Early Discovery of Coal, MS.

[57] Michael Shegda, "History of the Lehigh Coal and Navigation Company to 1840" (Doctor's thesis, Temple University, 1952) , pp. 37-43.

[58] Johnson, *Notes on the Use of Anthracite,* pp. 11-12.

liberate the carbon and impurities in order to make the iron malleable under hammer or roller, demanded skilled, experienced hands. It was one of the most important steps in the manufacture of finished iron products. In the United States, puddling or reverberatory furnaces burning bituminous coal had been used to purify pig iron since the introduction of the method in the western part of Pennsylvania as early as 1817. In furnaces of this type the iron did not come into direct contact with the coal; but there was a small degree of contamination from the sulphurous gases escaping from the fuel and absorbed by the melted iron.[59] The advantages of anthracite were obvious. Once ignited, the hard coal burned with a fierce heat, puddling pig iron in about forty-five minutes. There was little volatile matter in anthracite, thus reducing contamination from gases. Besides, it did not take as much anthracite to puddle a ton of pig metal as it did bituminous. Since western bituminous coal was more expensive in the East due to high transportation charges up to the mid-fifties, and foreign bituminous usually found a tariff barrier, eastern rolling mills, located on the anthracite trade lanes, welcomed the method of anthracite puddling. In 1827 Jonah and George Thompson, of Phoenixville, claimed to have puddled iron with anthracite coal.[60] In the early thirties experiments were carried on to perfect anthracite puddling. Another claim to the puddling process in which anthracite was used was made by Buckley and Swift, of Pottsville, in 1834.[61] The New Jersey Iron Company at Boonton also succeeded in puddling pig iron with anthracite coal in 1840.[62] But ironmasters did not consider anthracite puddling and reheating completely successful until 1844.[63] This was significant to eastern rolling mills, which found anthracite more economical than bituminous. The Montour Iron Works at Danville, for example, had sixty-nine anthracite puddling furnaces by 1856.[64]

Anthracite rolling mills, like anthracite blast furnaces, multiplied in eastern Pennsylvania, turning out iron rails for railroads, sheet iron,

---

[59] L. C. Hunter, "Influences of the Market upon Technique in the Iron Industry in Western Pennsylvania up to 1860," *Journal of Economic and Business History*, I (November, 1928-August, 1929), 246.

[60] J. M. Swank, *History of the Manufacture of Iron in All Ages* (Philadelphia, 1892), p. 364.

[61] Delaware and Hudson Minute Book, January 26, 1831; Hazard, *Register of Pennsylvania*, XIV (1835), 383; Hazard, *U.S. Register*, I, July 10, 1839, p. 33.

[62] *Ibid.*, II, August 19, 1840, p. 118.

[63] *Documents Relating to the Manufacture of Iron*, p. 66.

[64] *Pittsburgh Gazette*, January 9, 1857.

boiler plates, and various forms of bar iron. In 1847 there were six mills in Philadelphia.[65] Ten years later these mills had increased in size, added three to their number, and employed over seven hundred men.[66] Other establishments were found throughout the eastern part of the State. One of the largest, ranking with the Montour Works, was Reeves, Buck and Company at Phoenixville. Reading, Norristown, Harrisburg, Wilkes-Barre, Pine Grove, and other towns in the anthracite regions or on the transportation routes of the coal trade added to the growing number.[67]

At the same time, the number of firms manufacturing anthracite castings in cupola furnaces grew rapidly in Philadelphia, and supplied the needs of the stove industry for which the city became famous in the eighteen-fifties.[68] Schuylkill, Lehigh, and Lackawanna anthracite was used for castings in the ironworks and smith shops on the Great Lakes, the coal moving up the Hudson and over the Erie Canal. Anthracite was considered to be superior fuel for the cupola and was in demand to the extent that even the Tredegar Iron Works, of Virginia, pride of the South, used small quantities of Pennsylvania anthracite for that purpose. The coal was shipped by sea from Philadelphia in schooners which came south to take on cargoes of Virginia coal for the Philadelphia Gas Works.[69]

From blast furnace and cupola to forge hammer and rolling mill, hundreds of useful finished products were turned out. Anthracite iron had come of age in the East and threw its lengthening shadows of competition across the mountains to the valleys of western Pennsylvania. In the fifties considerable amounts of anthracite pig iron were being shipped from eastern Pennsylvania blast furnaces to supply the demands of the rolling mills and forges at Pittsburgh.[70]

COKE AND BITUMINOUS

The application of anthracite coal to the iron industry gained almost spectacular success in eastern Pennsylvania and in regions

---

[65] *Merchants' Magazine*, XVI, 593.

[66] Freedley, *Philadelphia and Its Manufactures*, p. 287.

[67] *Merchants' Magazine*, XVI, 593.

[68] Freedley, *Philadelphia and Its Manufactures*, pp. 289-291.

[69] Taylor, *Statistics of Coal*, p. 312; Kathleen Bruce, *Virginia Iron Manufacture in the Slave Era* (New York, 1931), p. 212.

[70] Wetherill, *Report on the Iron and Coal of Pennsylvania; Pittsburgh Quarterly Trade Circular*, p. 40; CSHS, Palmer Papers, notes on an interview with Mr. Christopher of Pittsburgh, August 20, 1856,

adjacent to anthracite coal supplies. In the western part of the State, near Pittsburgh, bituminous coal deposits had been worked for years to furnish fuel for the city's homes and industries, and especially for her rolling mills and forges. Bituminous coal was excellent for heating and puddling pig iron. There was no direct contact between metal and fuel in the process, and therefore the impurities mattered little. Yet neither raw bituminous coal nor coke had been applied with telling effect to the blast furnaces of western Pennsylvania before the decades of the thirties and forties. For several years, even after success had been achieved, coke and bituminous pig iron were not well received by western ironmasters or consumers.

There were several reasons why the use of coke and raw bituminous coal in the blast furnaces of the West before 1860 faltered and lagged behind charcoal and the eastern rival, anthracite. Some of these reasons are found in the technological problems of the coking process, the proper blast for the iron furnaces, and the need for better grades of coking coal. Then too, the presence of almost limitless supplies of timber for charcoal from the great parallel ranges of the Alleghenies strengthened the prejudice of many ironmasters and consumers in favor of the tough, malleable charcoal iron which met the demands of the western agrarian market. That some charcoal ironmasters were against the manufacture of coke pig iron was evident during the debates in the State legislature in 1831-32 over the passage of the act to incorporate the Pennsylvania Coke and Iron Company. Charges of monopoly and lobbying were made against the charcoal ironmasters by the *Philadelphia Gazette* and Condy Raguet's free trade journal, the *Banner of the Constitution*. Defeated in the lower house in 1831, the bill was passed the following year by the narrow vote of fifty-one to forty-six. Although unalterably opposed to the incorporation of companies for business which properly belonged to individuals, the *Banner of the Constitution* hailed the incorporation of the coke iron company as a telling blow against the charcoal iron interests and the beginning of an era of cheap iron through cheap fuel.[71] Raguet, in the zeal of a crusader, was carried away by the cause of free trade. A law can protect and encourage, but it cannot establish a manufacturing process nor solve its technical problems. It would be many years before cheap coke iron would supersede the more expensive charcoal iron.

---

[71] *Banner of the Constitution,* February 22, March 14, 1832.

The main object of coking was to free bituminous coal from impurities, particularly sulphur, so that it could be used in smelting iron. To achieve this, coal was burned in piles or rows on a level space or yard surrounded by a ditch filled with water. This area was called the coking pit. At the start of the process the coal was burned slowly in a moist, smoldering heat which drove off the other impurities, leaving a heavy, silvery white coke high in carbon. When a high, dry heat was applied at the start, the carbureted hydrogen gas was driven off first, but the sulphur remained and combined with the carbon.[72] The result was a poor grade of coke unfit for the blast furnace. Later in the period, coking ovens of the beehive type were erected and the process speeded by regulating the heat with a greater degree of accuracy. Later the escaping gases were utilized as fuel for the ovens.[73]

The coking process, perfected in England, was not well understood in eighteenth-century America nor in the early days of the federal union. Still, some of the English and Welsh ironworkers migrating to the new United States after the Revolution undoubtedly knew something about the manufacture of coke. One of these artisans was John Beal, who published an advertisement in the *Pittsburgh Mercury* in 1813:

> To Proprietors of blast-furnaces:
> John Beal, lately from England, being informed that all the blast furnaces are in the habit of smelting iron ore with charcoal, and knowing the past disadvantages it is to proprietors, is induced to offer his services to instruct them in the method of converting stone coal into coak. The advantages of using coak will be so great that it cannot fail becoming general if put to practice. He flatters himself that he has had all the experience that is necessary in the above branch to give satisfaction to those who feel induced to alter their mode of smelting their ore.
>
> John Beal, Iron Founder.
>
> N.B. A line directed to the subscriber will be duly attended to.[74]

---

[72] J. P. Lesley, *Manual of Coal and Its Topography* (Philadelphia, 1856), pp. 27-29; see also, John Fulton, *Coke* (Scranton, 1905).

[73] CSHS, Palmer Papers, Specifications of Patent to John Cox-Patent Coke Oven 1840-41; Coking at Hollidaysburg, Pennsylvania, Gardiner and Company Furnaces. July, 1857.

[74] *Pittsburgh Mercury*, May 27, 1813. Reprinted in J. D. Weeks, *Report on the Manufacture of Coke, Tenth Census of the United States*, X (1880), 23.

No one knows if Beal's offer was accepted, but it seems improbable that it could have been. There was no blast furnace in the immediate vicinity of Pittsburgh at that time, nor was there one for nearly half a century.

In 1816-17 Colonel Isaac Meason built the first rolling mill west of the Alleghenies at Plumsock in Fayette County, Pennsylvania. Here, for the first time in the United States, so far as records indicate, coke was made and put to use in puddling and heating iron. The man responsible for this was an employee of Meason, a Welsh ironworker, Thomas C. Lewis. The first coke blast furnace in the United States, the Bear Creek Furnace in Armstrong County, Pennsylvania, was designed and put into operation by this same Lewis in 1819. The blast was too weak, however, and after producing a ton or two of coke pig, the furnace chilled and went out.[75] But the failure to make coke pig iron did not result in the abandonment of the furnace. For many years it was used to produce charcoal iron.

Scattered experiments with coke continued without success. A few enterprising Pennsylvanians comprising the Pennsylvania Society for the Promotion of Internal Improvements despaired that although bituminous coal and iron ore abounded in western Pennsylvania, the lack of technical information on the manufacture of coke and its application to the blast furnace made it impossible for the State to capitalize on the real value of these resources. "Attempts of the most costly kind have been made to use the coal of the western part of our state in the production of iron," said the society, and went on to relate that "furnaces have been constructed according to the plan said to be adopted in Wales and elsewhere; persons claiming experience in the business have been employed, but all has been unsuccessful."[76] Inspired by British methods in the field, the society sent William Strickland to Great Britain in 1825, allotting him £100 to investigate the latest procedures of coke manufacture and the use of coke in the blast furnace.

Strickland's report lent some encouragement to further experimentation. By 1834 the Packer Committee, reporting on the coal trade to the Senate of Pennsylvania, stated, "The coking process is now understood, and our bituminous coal is quite as susceptible of this operation, and produces as good coke as that of Great Britain.

---

[75] *Tenth Census*, X, 23.
[76] *Ibid.*

. . . There is nothing to prevent us from becoming a great and powerful manufacturing people."[77]

The fundamentals of the coking process may have been understood, but there still was room for considerable improvement twenty years later. Evidence of this was shown in the interest displayed in British coking methods about 1855 by the Pennsylvania Railroad.[78] The ardent faith in American methods and in the future of Pennsylvania resources was a part of the era, and we should not be too critical of the zealots of the "Coal Age." But the report failed to mention that in 1834 coke had not been applied successfully to the blast furnace. Encouragement by the Franklin Institute and the Pennsylvania legislature for the successful manufacture of iron with coke or mineral coal in 1835 and 1836 showed that western blast furnaces still relied on charcoal. Between 1835 and 1839, however, there were several Pennsylvania experiments with coke in blast furnaces which approached success, but did not attain it.

Broad Top coal of Huntingdon County was coked and used in the Mary Ann Furnace owned by William Firmstone, who claimed to have made good iron for one month during 1835.[79] The next year one hundred tons of coke pig iron were produced by the Oliphants at their Fairchance Furnace near Uniontown, Fayette County. The Oliphants abandoned their experiments due to the poor response of the coke iron under the forge hammer and returned to charcoal.[80] Peter Ritner and John Say at Karthaus on the West Branch of the Susquehanna, under the name of the Clearfield Coke and Iron Company, made coke pig iron in 1838.[81] Henry C. Carey, John White, Burd Patterson, and others bought out the company and employed William Firmstone, who used a hot blast. But poor transportation and inferior ore rather than poor coking coal put an end to the project the next year.[82] About the same time experiments were carried on farther east at Farrandsville, Pennsylvania, where several hundred tons of coke pig iron were made between 1837 and 1839.

---

[77] *Pennsylvania Senate Journal*, 1833-34, II, 483; for another optimistic view on the subject see *Essay on the Manufacture of Iron with Coke* (Lewistown, Pa., 1838).

[78] CSHS, Palmer Papers, W. J. Palmer to J. Edgar Thompson, September 3, 1855.

[79] *Tenth Census*, X, 24.

[80] Bining, "The Rise of Iron Manufacture in Western Pennsylvania," pp. 247-248.

[81] *Pennsylvania House Journal*, 1837-38, I, pp. 542, 855; W. R. Johnson, *Report on an Examination of the Mines, Iron Works, And Other Property Belonging to the Clearfield Coke and Iron Company* (Philadelphia, 1839), pp. 1-12.

[82] Johnson, *Notes on the Use of Anthracite*, p. 6.

At Farrandsville, it will be recalled, Benjamin Perry, familiar with hot-blast methods used in his native Wales, not only ran off coke pig, but also experimented with raw bituminous coal in the blast furnace. This was before Burd Patterson hired him to take charge of the anthracite furnace at Pottsville.[83] Perry, who distinguished himself in the anthracite iron industry, was unsuccessful in his coke and raw bituminous experiments. W. R. Johnson, an outstanding contemporary authority, blamed the inferior coal of the region and not the technological skill of the furnace master.[84]

While Pennsylvania ironmasters tried valiantly to make a good coke pig iron, the George's Creek Coal Company built the Lonaconing Furnace near Frostburg in western Maryland. Coke made from the Cumberland bituminous coal proved to be of fair quality. By 1839 the furnace was producing seventy tons of hot-blast coke pig iron a week.[85] In the same state the Mount Savage Company built two furnaces in 1840 and used the hot blast to produce between five and seven thousand tons of coke pig per year.[86]

By the eighteen-forties a fair grade of coke could be produced. Still, in that decade there were only four coke furnaces in blast in Pennsylvania. These belonged to the Western Iron Works at Brady's Bend on the Allegheny River, about forty miles north of Pittsburgh. Most of the Pittsburgh manufacturers who turned out iron products for the agrarian markets of the West refused to use coke pig iron, which, for their purposes, was inferior to charcoal pig. The Western Iron Works, therefore, erected its own rolling mill to process the iron from its furnaces.[87]   But inferior coking coal containing large amounts of sulphur, coupled with inexperience in making coke for the blast furnace, proved to the company the prejudice of the Pittsburgh mill owners. The "red-short" iron, brittle when hot, did not fit the needs of the market.[88]

It was in the same decade, however, that the Connellsville sector of the great Pittsburgh seam was discovered and opened for development. The coal found in this narrow, sixty-mile trough drained by

[83] Hazard, *U.S. Register,* I, July 3, 1839, p. 27.
[84] Johnson, *Notes on the Use of Anthracite,* pp. 7-8; Swank, *Iron Making and Coal Mining in Pennsylvania,* p. 75.
[85] Bining, "The Rise of Iron Manufacture in Western Pennsylvania," p. 248.
[86] *Tenth Census,* X, 25.
[87] Bining, "The Rise of Iron Manufacture in Western Pennsylvania," pp. 248, 250-51.
[88] A. C. Bining, "Ironmen in Quest of Fuel," *Steelways,* X, August, 1954, p. 11.

the Youghiogheny River produced a superior coke. Connellsville coke
was of a silvery lustre and cellular, with a metallic ring. Above all,
it was comparatively free from impurities and possessed a strong
structure capable of bearing the heavy burden of the blast furnace
charge. Purity and strength made it the best blast furnace coke in
the United States.[89] Its manufacture began in 1841-42, but because
of the persistence of the western agrarian market and the lean years
in the coal and iron industry during the latter part of the forties, the
development of the Connellsville coke area did not make real progress
until the next decade. The utilization of Pennsylvania Connellsville
coke as a blast furnace fuel was the true beginning of the coke iron
industry.

The number of coke blast furnaces grew slowly in Pennsylvania.
The years between 1846 and 1850 were marked by a depression in the
iron industry. Ironmasters and coal operators, who felt the economic
pinch, blamed bad times on the low tariff policies of the Polk Administration and increased competition with foreign products, especially Scotch pig iron and British coal. Depression took the toll
of the few early coke blast furnaces. In 1849 there was not a single
one in operation in the State. Recovery came early in the fifties, and
by 1856, with better times and new demands, there were twenty-one
coke blast furnaces producing a total of almost forty thousand tons
of pig a year.[90] This amount, of course, was only a fraction of the
anthracite iron of the East and amounted to less than half the
charcoal pig of the Commonwealth. The production of iron from
raw bituminous coal was even smaller, with its tonnage under nine
thousand.[91]

The effects of the depression following the Panic of 1857 were soon
shaken and the iron and coal industries continued to expand. With
the exception of this brief interlude, the changing economy of the
West was reflected in the slow but steady increase of coke and bituminous pig iron. The fact that the United States still imported half
its iron rails from England indicated a growing need for cheaper
methods of iron manufacture to meet the demands of railroads and
industry. There was tremendous promise in the new coke iron. This
was shown shortly before the Civil War, not only in the comparative
cheapness of the metal, but in the expansion of the Connellsville

---

[89] F. E. Saward (ed.), *The Coal Trade* (1878), p. 19.
[90] *Tenth Census,* X, 26.
[91] See Table D, Appendix.

coke area. Still, the challenge of eastern anthracite iron had not been met nor the practical virtues of charcoal iron superseded by western coke iron manufacture by 1860.

Charcoal pig iron, though expensive, possessed qualities important to the rural economy. When forged or rolled into wrought iron it was malleable both hot or cold. In the forge of the rural blacksmiths it could be shaped into implements for farmers, wagoners, and mill owners who peopled the western river valleys. Large amounts of the charcoal pig iron and blooms puddled, heated, and rolled in the great mills at Pittsburgh were sent to the rural craftsmen, the blacksmiths, who made iron articles for agrarian consumers. In the fifties, the economy of the West, like the economy of the East a decade before, began to show marked changes in the demand for iron products. Industries and railroads needed boiler plates, machinery, castings, and iron rails. These could be made from the cheaper but less malleable product of the coke furnaces. By the fifties coke iron, while not superior to charcoal iron, was smelted, puddled, heated, and rolled with satisfaction and was considered good iron for industrial use.[92]

When the changing market could absorb iron made from coke as well as from raw bituminous or bituminous mixed with coke, cheap mineral fuel drew many ironmasters and capitalists into a position to profit from the manufacture of the new iron.[93] In the West, mountain timberland could be purchased at low prices as the slopes could not be used as profitable farm land. But charcoal in this wild country was expensive. Considerable labor was demanded to cut, haul, and char the forest wood for the blast furnaces. Figures are difficult to obtain for the early period, but in the eighteen-fifties western charcoal, like eastern charcoal, was priced as high as five cents per bushel. It took 180 to 200 bushels to smelt one ton of charcoal pig iron at a fuel cost of $8.50 to $10.00. The Cambria Iron Works near Johnstown made charcoal pig in four furnaces in 1849-50. In 1850 it built four coke blast furnaces.[94] In 1853-54 these works were using coke fuel at fifty-four cents per ton, averaging five tons of coke for each ton of coke pig iron. The Lycoming Iron and Coal Company at the same time set the fuel figures even lower: $2.10 per ton of coke pig

[92] Hunter, "Influence of the Market upon Technique in the Iron Industry in Western Pennsylvania up to 1860," pp. 263-264.

[93] Pennsylvania Legislative Documents, 1853-1854, pp. 311-312; Taylor, Statistics of Coal, pp. 422-423.

[94] Wetherill, Report; Taylor, Statistics of Coal, pp. 422-423.

to $6.38 per ton of charcoal pig.[95] Not all works using bituminous coal, either in raw form, mixed, or as coke, were as fortunate, and fuel prices were higher if local coal was not used. The Gardiner and Company furnaces, in 1857, paid five cents per bushel or $1.40 per ton for bituminous coal hauled over the Allegheny Portage Railroad. At the ironworks it was made into coke, both in open pits and in ovens. There were twenty-five ovens in operation on the company's premises, each holding fifty bushels of coal and coking it in twenty-four hours. The open pit process was much slower. Six to ten tons of coal were burned six to eight days, then watered and dried for eight hours. The time and inefficiency involved in this latter process made the coke oven the preferred method.[96]

Superior Connellsville coke was shipped to the ironworks at Pittsburgh where it was used in rolling mills and forges. In 1859 the Clinton Furnace of Graft, Bennett and Company, the first blast furnace at Pittsburgh, was blown in. During the first three months the furnace used coke made from the coal beds near the city. This coke proved unsatisfactory because of excessive impurities. The following spring coke was hauled over the Baltimore and Ohio Railroad from the Connellsville area and proved to be as excellent as its reputation.[97]

The coke pig iron produced by the Clinton Furnace added to the supplies of "raw iron" processed by the great mills and forges of the city. The expected saving in fuel was offset by the cost of ore which was not obtained in the vicinity. Pigs and blooms continued to move to Pittsburgh from various sources. Charcoal pig iron and blooms from Tennessee and the valleys of the Juniata, Allegheny, and Monongahela brought an average price of thirty dollars per ton on the Pittsburgh market. This was undersold by anthracite iron from the eastern part of the State which could be purchased for $25 a ton. Coke iron and small amounts of pig smelted by raw bituminous coal drawn from the surrounding counties of Lawrence, Mercer, Beaver, Fayette, and Cambria were priced between seventeen and eighteen dollars.[98]

---

[95] Hunter, "Influence of the Market upon Technique in the Iron Industry in Western Pennsylvaina up to 1860," pp. 263-264.

[96] CSHS, Palmer Papers, Coking at Hollidaysburg, Gardiner and Company Furnaces, July, 1857.

[97] *Tenth Census*, X, 26-27.

[98] *Pittsburgh Quarterly Trade Circular*, I, 40-42.

The raw bituminous pig iron industry, turning out scarcely nine thousand tons in 1856, was to be found in the Mahoning and Shenango valleys to the northwest. The method of smelting with raw bituminous first succeeded in Ohio and spread to Pennsylvania in the forties, but the iron was not considered to be as good as coke iron. Much of the raw bituminous coal used in experiments crumbled under the pressure of the charge as it was too soft to stand the weight of the ore and flux. Coke, on the other hand, if made properly, was solid enough to withstand the weight. The bituminous coal found in the Shenango Valley in Mercer County and along the Mahoning which rises in Ohio was of different structure than most Pennsylvania bituminous. It was laminated, splitting into flat blocks, and was very hard to break across the laminations.[99] "Free-burning splint" or "block coal," as it was called, did not make solid coke because of its structure.[100] But in the raw state, because of its strong structure, it could bear the weight of the blast furnace charge. Fortunately the coal contained little sulphur. Like Connellsville coke, block coal possessed purity and strength. Under the hot blast these qualities permitted satisfactory utilization in the raw state as a smelting fuel.

The use of coke and bituminous coal in the blast furnace increased the consumption of Pennsylvania coal approximately a quarter of a million tons a year for the five years preceding the Civil War. Large amounts of the coke consumed came from the Connellsville region. Unlike anthracite, a Pennsylvania monopoly, coke or bituminous used in the blast furnaces outside the State usually was obtained in the vicinity of the furnaces. The annual quarter of a million tons consumed by blast furnaces in Pennsylvania was equaled by the rolling mills and forges of Pittsburgh. Bituminous coal maintained its supremacy in the manufacture of finished iron products. Despite use in the blast furnace of coke made from bituminous coal, bituminous had failed by 1860 to surpass its anthracite or charcoal competitors.

[99] Eavenson, *American Coal Industry*, p. 225.
[100] Swank, *Iron Making and Coal Mining in Pennsylvania,* p. 77.

CHAPTER V

Steam Vessels

STEAMBOATS ON INLAND WATERS

FOR more than thirty years before the Civil War experiments were carried on in the field of coal-powered steam navigation. Wood-burning steam vessels plying the vast network of inland waterways in the United States were viewed by both anthracite and bituminous coal producers as important potential customers of mineral fuel. In the East there was considerable technical success in the adaptation of marine boilers to anthracite coal. Eastern river and sound steamers, by the forties, were using large amounts of Pennsylvania anthracite. Steamboats on the river routes of the western country did not make a complete transition from wood to mineral fuel as did many of the eastern vessels, despite fewer technological problems in burning bituminous under marine boilers.

Coal was really the first steamboat fuel on western rivers, as it was burned by the "New Orleans" on part of her maiden voyage in 1811-12. Her owner and promoter, the young, ambitious Nicholas J. Roosevelt, of New York, had carefully surveyed the Ohio-Mississippi River road in 1809. He also collected a quantity of Ohio coal near Pomeroy, a point on the Ohio River about halfway between Pittsburgh and Cincinnati, to be used in refueling his steamboat which he hoped to build and launch as soon as he returned to Pittsburgh.[1] The "New Orleans" burned some Pittsburgh and Pomeroy coal, but she also burned quantities of wood, for coal supplies along the Ohio in 1811-12, with the exception of Pittsburgh and the cache at Pomeroy, were practically nonexistent. Many steamboats followed the "New Orleans" on the Ohio, Mississippi, and Missouri, the three great river arteries of the West. But the common use of bituminous coal in western steamboats was hampered and delayed by a number of practical problems which were not faced by the eastern vessels.

River steamboats in the West rarely carried wood supplies for more than one day's run. The chief consideration was the weight saved by lighter fuel loads and not so much the amount of space conserved.

---

[1] C. H. Ambler, *A History of Transportation in the Ohio Valley* (Glendale, California, 1932), pp. 113, 121.

Sand bars, low water, half-submerged driftwood islands, and other hazards of western river navigation made it imperative that the steamboats maintain a shallow draught. In the early days, when population was sparse, the usual practice of the steamboat captains was to send crews ashore to cut wood from the timberlands near the banks of the river. As the population increased along the Ohio, Mississippi, and Missouri, the adjoining timberland was claimed by settlers, and the cutting of wood for steamboat fuel became an important industry for people living along the riverbanks.[2] This was more satisfactory than before, as the wood usually was seasoned and burned better than the green timber secured by boat crews. The best wood was resinous pine found along the lower Mississippi and was much preferred by boat captains when passing through that section of country. Still, the masters of the vessels had to be on the alert for sly riverbank merchants who would sell them green timber of any kind if they had the chance. A story reflecting this particular situation appeared in a Cincinnati paper in 1845. A Mississippi steamboat captain called to a wood merchant on shore, "What wood is that?" "Cord wood," came back the answer. "How long has it been cut?" asked the captain. "About four feet," replied the wood merchant.[3]

The problem of less wood cargo and suitable wood fuel was complicated further by the increased costs of wood in the twenty years before 1860. Timber became less plentiful along the banks of the Ohio and central Mississippi as the forests retreated before the axe, the farm, and the town. Steamboats placed an additional burden on the forested areas. Small and medium-sized river steamers burned from twelve to twenty-four cords of wood a day, and the large boats consumed as much as fifty to seventy-five cords for every twenty-four hours running time.[4] Bituminous coal, most of which was mined in Pennsylvania, had become a welcome substitute for wood fuel in the factories, mills, and homes of Cincinnati, Louisville, and St. Louis. By the fifties most of the Ohio and some of the Mississippi steamboats burned bituminous coal, but mineral fuel did not completely supplant wood under the boilers of the riverboats.

Coal, like wood, varied in price along the meandering water routes of the Mississippi and Ohio. At the headwaters of the Ohio at Pittsburgh, steamboats standing downriver usually carried a day's

---

[2] L. C. Hunter, *Steamboats on the Western Rivers, an Economic and Technological History* (Cambridge, 1949), pp. 264-265.

[3] *Cist's Weekly Advertiser*, October 15, 1845.

[4] Hunter, *Steamboats*, p. 266.

supply of Pennsylvania bituminous, which could be purchased at the Pittsburgh wharves for a few cents a bushel. Approximately 150 bushels of coal could supply the fires of a medium-sized steamboat for one day's run. A heavier fuel cargo was the exception as it increased the draught of the vessel.[5] In the thirties the western Pennsylvania coal trade was not too well developed. Coal supplies on the lower Ohio, even at Cincinnati and Louisville, were scarce and expensive; thus much of the Ohio run was made on wood.

During the next two decades, from 1840 to 1860, the coal trade from the Pittsburgh and Monongahela regions increased. The development of coal mines along the Ohio and its lower tributaries added to available coal supplies in the valley. Coal still was scarce on the lower Mississippi and little could be found on the banks of the Missouri.

John Randolph of Roanoke is said to have described the Ohio River as "frozen up" one half of the year and "dried up" the other half. While not conforming to the description, the great river, flowing nearly a thousand miles from Pittsburgh to its junction with the Mississippi at the southern tip of Illinois, was subject to the vagaries of weather. Ice and drought hampered Ohio River traffic at times and influenced the supply of coal available to steamboats. In periods of low water the small, shallow-draught steamboats could leave the wharves of Pittsburgh and other Ohio River ports before the "coal boat rise." However, coal boats were great box-like affairs, as bulky as the cargo they carried. Loaded to the gunwales, the coal boats not only drew considerable water, but needed high water for safe handling. Floating ice was another hazard for the coal boats, which were difficult enough to maneuver when the river was ice free.[6] Steamers continued to use the water routes though river towns had not replenished their coal supplies. Wood fuel, though expensive, at least was available, and until coal, the cheaper fuel, was again obtainable, wood was used.

Because of the scarcity or the uncertainty of bituminous supplies on the rivers of the western country, steamboats were seldom equipped to burn coal as their only fuel. Bituminous coal required a smaller firebox and finer grate bars than wood to insure efficient combustion.[7]

---

[5] *Ibid.,* p. 267.

[6] L. C. Hunter, "Seasonal Aspects of Industry and Commerce Before the Age of Big Business," *Smith College Studies in History,* XIX (October, 1933-July, 1934), p. 15.

[7] *Cannelton, Perry County, Indiana,* pp. 80-81.

The standard procedure was to compromise with a medium-sized firebox which sacrificed combustion efficiency, but was large enough to burn wood as well as coal. Most Ohio River steamboats burned a mixture, soft coal and wood. The coal was kept at the base of the fire and the wood scattered on top. Furnaces were fired by alternate layers of coal and wood. A very hot fire was the result and steam could be produced quickly. This was always welcome in the frequent emergencies of river navigation. Many steamboat engineers often preferred wood to soft coal, despite the expense of the former, as it was a clean fuel and did not necessitate frequent cleaning of the flues.[8] Also, bituminous coal soot pouring from the stacks of the steamboats was particularly annoying to the patrons of the passenger packets.

For years western coal producers tried to convince steamboat operators that bituminous fuel should be used exclusively. Cost, space, and weight were stressed as advantages of coal over wood. But the problem of dependable supply remained, and wood continued to supplement mineral fuel in western steamboats.[9]

As late as the mid-fifties coal firms were complaining that while western steamboats used bituminous coal, wood was still an important part of their fuel.[10] In 1855, Wheeling, Virginia, an important coal center, opposing the railroad between Baltimore and Columbus, argued that she would become a "mere wooding station for steamboats" should the proposed railroad bypass her.[11]

Pennsylvania, Ohio, and some Illinois coals were introduced on the Great Lakes in the late forties. Small quantities of Pennsylvania anthracite and Blossburg semibituminous found their way to Buffalo. Larger amounts of bituminous coal came to Lake Erie from Mercer and Lawrence counties in northwestern Pennsylvania. Canal, and later railroad, extensions brought Pennsylvania coal to the cities of Erie and Cleveland. A major consideration in the construction of both the Erie extension of the Pennsylvania Canal and the Cleveland

---

[8] Hunter, *Steamboats*, p. 266.

[9] *Cist's Weekly Advertiser*, August 1, 1851; "The fuel used in the *Magnolia* is, when running up stream, and in the cotton trade, yellow pine; and down stream cottonwood ash and cypress; when running down in the Louisville and New Orleans produce trade, Pittsburgh coal is used, with the yellow pine wood . . . ," *Journal of the Franklin Institute,* LVI, April, 1853, p. 259.

[10] CSHS, Palmer Papers, Prospectus of the Pittsburgh and Youghiogheny Coal Company (rough copy), 1856.

[11] Ambler, *Transportation in the Ohio Valley,* p. 203.

and Pittsburgh Railroad was the coal trade, with particular emphasis on the potential coal consumption of Lake steamers.[12] Steamboats on the Great Lakes were using large amounts of Pennsylvania and Ohio coals by the mid-fifties. These craft readily adopted bituminous coal as fuel, because, unlike the river steamers of the West, they had little concern for shallow draught. On the Lakes, not weight but space and steaming range were the major considerations in taking on fuel. One ton of coal was equivalent to over three cords of wood and took up much less space. Additional supplies gave the steamers greater range in their trips across these inland seas. Lake steamers burned huge amounts of wood, and the cost of wood was high. The *Chicago Tribune*, in 1848, reported that the steamboat "Empire" burned seven hundred cords of wood between Chicago and Buffalo. The boat made thirteen trips per season which meant that forty wood cutters destroyed two hundred and thirty-four acres of timber at a cost of ten thousand dollars for wood and wages. In 1848 there were fifteen other first-rate steamers on the upper Lakes.[13] Wholesale destruction of timberland and mounting costs beckoned the bituminous coal trade of Pennsylvania. The price of bituminous coal at Cleveland and Erie averaged between $2.50 and $3.00 per ton, while at Buffalo it was $4.00 per ton. Small shipments of Illinois coal found along the Illinois and Michigan Canal began coming into Chicago in 1849-50. The Illinois bituminous did not compete with Pennsylvania or Ohio coals in steam navigation as the sulphur content was so high that grates and boilers suffered damage. Illinois coal was considered useless by steamboat engineers.[14]

Pennsylvania coal also found a market in the Canadian Lake ports and was used by some Canadian steamers. In 1853 the *Cleveland Herald* estimated the entire consumption of the Lake region to be three hundred thousand tons.[15] More than half of this amount came from Pennsylvania. The mines in Mercer County that year shipped over one hundred thousand tons to Erie. Lawrence County the year before had started small shipments amounting to ten thousand tons, and additional supplies reached Cleveland via the Cleveland and Pittsburgh Railroad. Within the decade the Pittsburgh, Fort Wayne

[12] W. M. Roberts, *Report on the Erie Extension to the Board of Canal Commissioners, Pennsylvania House Journal*, Appendix to II (1840), pp. 273-274; see also *Annual Report of the Cleveland and Pittsburgh Railroad, 1849-1856*.

[13] *Niles' National Register*, LXXIV (July-December, 1848), 394.

[14] Taylor, *Statistics on Coal*, pp. 312-313.

[15] *Ibid.*, p. 313.

and Chicago Railroad pushed through to Lake Michigan and served as a connecting link between the Pittsburgh-Monongahela coal fields and the Lakes. By 1860, Pennsylvania bituminous coal shipments to Lake ports had increased to more than a quarter of a million tons a year.[16] A conservative estimate would be that one-third of this amount was burned as fuel by Lake steamers. But like the river steamboats of the West, the Lake steamers still supplemented their coal fuel with wood, primarily because of the problem of inadequate and unreliable supplies of mineral fuel at Lake ports.

The eastern portions of the country touched by the Pennsylvania anthracite trade showed an early interest in the use of hard coal as fuel for river and sound steamers. Anthracite coal firms led the way in the encouragement of anthracite steam navigation. The Lehigh trade was only six years old in 1826 when the Lehigh Coal and Navigation Company began experimental trials with anthracite in the steam boilers of towboats for their coal barges.[17] In 1831 the Lehigh company purchased a wood-burning steamer, the "Pennsylvania," for further experiments. For less than one hundred dollars the grates of the firebox were altered to burn coal and the boat was put into operation. For several years the "Pennsylvania," using anthracite fuel, towed coal arks on the Delaware River between Philadelphia and Coal Haven near Trenton. It took the small steamboat nearly an hour to get up steam for the thirty-three-mile trip, but the amount of fuel saved by using anthracite instead of wood made up for the slowness in producing steam.[18] The length of time it took to raise steam aboard the "Pennsylvania" indicates that the technology of anthracite marine boilers was in its infancy. The steamboat used three tons of coal per round trip of sixty-six miles at half the cost of wood, and could tow three sets of Delaware coal arks carrying a total coal load of four hundred tons.[19] The "Pennsylvania" soon had a "sister ship," the "Convoy." By 1839 the Board of Managers of the Lehigh Coal and Navigation Company reported to the stockholders with satisfaction on the gradual introduction of their anthracite for steamboat fuel on the Delaware, the Hudson, and Long Island Sound.[20]

---

[16] Eavenson, *American Coal Industry*, production tables, pp. 464, 472, 491, 493, 496.

[17] HSP, Josiah White to Governor Joseph Ritner, July 11, 1837.

[18] Hazard, *Register of Pennsylvania*, VIII, 15; XIV, 144.

[19] *Ibid.*

[20] *Report of the Board of Managers of the Lehigh Coal and Navigation Company,* January 14, 1839, p. 32.

The dozen years between the first experiments with anthracite in eastern steamboats and the Lehigh company's report were marked by concern among steamboat operators over the growing scarcity and rising cost of wood. This concern was reflected in the many attempts to adopt anthracite fuel in marine boilers and insure maximum operating efficiency. New York steamboats consumed an estimated two hundred thousand cords of pine wood during an eight-month running season in 1828. Philadelphia's steamboats, ferries, and factories used one hundred and fifty thousand cords.[21] Some of the supply came from the pine lands of the Carolinas, but most of it was cut from the shrinking pine barrens of southern New Jersey. At the turn of the century in 1800, before the advent of the steam engine in manufacturing and transport, Jersey pine lands, then considered unfit for agriculture, were worth six to ten cents per acre. Steam revolutionized the land value of this wooded region, which was close to Philadelphia and, by sea, not far from New York. Within a generation the price of an acre of pine timber had risen to six dollars and threatened to go higher as labor costs mounted and the demand for quick-burning, resinous pine wood increased with factory and steamboat development.[22] Steamboat men, like factory owners employing stationary steam engines, looked about for a cheaper substitute fuel.

The stationary steam engine began using anthracite coal in 1825. But with the exception of the "Pennsylvania" and the "Convoy," which operated on short runs, and then not too efficiently, and scattered experiments in New York, steamboat engines lagged behind the stationary engines in the use of anthracite fuel for a decade or more. The steamboat engine was more powerful and more complex than the stationary engine. While the stationary engine was simple to operate and maintain, running at a fairly uniform rate with few excessive pressures, steamboat machinery often was overtaxed and had to react quickly to signals from the bridge. It was imperative that steam pressure be maintained, controlled, and altered during a run in order to maneuver the boat. Thus, a strong, hot fire under the boilers was necessary at all times. When anthracite was tried in boats that burned wood, it proved inefficient. The flames from the coal did not reach high enough from the deep firebox to have telling effect on the boilers. This was the basic problem in stationary engines, but it

---

[21] *Niles' Weekly Register*, XXXIV (March-August, 1828) , p. 362.

[22] *Philadelphia Saturday Bulletin*. Reprinted in the *Miners' Journal*, August 1, 1829.

was a greater problem in steamboats due to the demands for quick, ready steam at all times. The problem of reliable and adequate coal supplies was voiced by the Pottsville *Miners' Journal* in 1827 as being the only reason why anthracite had not been adopted by steamboats. The argument was invalid.[23] Eastern river and Long Island Sound steamers did not adopt anthracite as their common fuel until 1838-40. Adequate supplies of Pennsylvania anthracite had been available along the Delaware and Hudson rivers and in the coastal towns and cities from Baltimore to Boston since 1833. The delay in the use of anthracite in eastern steam vessels was not so much a problem of supply as a problem of technology.

The leader in the large-scale introduction of anthracite fuel for steamboats, particularly those on the Hudson River and Long Island Sound, was the Delaware and Hudson Company. The company mined and shipped Pennsylvania anthracite from the Lackawanna Valley to the port of New York via the Delaware and Hudson Canal and the Hudson River. Lackawanna coal ignited more easily than the heavier anthracite of the Lehigh and Schuylkill areas. From the beginning of the first shipments, the company believed that its comparatively light anthracite was excellent steam coal for stationary engines, and told stockholders that Lackawanna coal "will, ere long become the most favored article for the same purpose in steamboats."[24] Little was accomplished in the introduction of Lackawanna coal aboard steamboats until 1831 when the company noted that three ferry concerns on the East River and one on the Hudson used Lackawanna with "entire success," and two or three coastwise steamers out of New York burned its fuel.[25] The next year a new vessel, the "David Brown," was constructed to burn Lackawanna coal. It was hoped that the boat could make the passage from New York to Charleston without touching any intermediate port to refuel.[26] The Delaware and Hudson Company realized that improvements were necessary in anthracite steamboat engines if eastern vessels were to adopt Lackawanna coal in place of pine wood, and that the phrase, "entire success," had been premature. The company, at its expense, altered the fireboxes

---

[23] *Miners' Journal*, December 1, 1827. For another misleading statement concerning the unavailable supplies of Pennsylvania anthracite for steam vessels see D. B. Tyler, *Steam Conquers the Atlantic* (New York, 1939), p. 128.

[24] *Annual Report, Delaware and Hudson Company* (1828), p. 4.

[25] *Ibid.*, (1832) p. 5-6.

[26] *New York Evening Post.* Reprinted in J. C. Emmerson, Jr. (ed.), *Steam Navigation in Virginia and Northeastern North Carolina Waters, 1826-1836* (Portsmouth, Virginia, 1949), p. 217.

and grates in one of the East River ferries and sent free coal for experiment to the Walnut Street Ferry in the spring of 1831.[27] The board of managers also delivered coal to the steamboat "Victory" for trial runs between New York and Hartford.[28] The company's untiring efforts to promote the use of Lackawanna anthracite in steamboats became well known among the steamboat men of the eastern coast. Once the board of managers was approached by a Mr. J. P. Allane with the proposition that the Delaware and Hudson take stock, payable in coal, in a new steamboat which was to run from Boston to Bangor. The board deliberated the proposal, but finally declined on the grounds that it did not possess the power to subscribe to the stock.[29]

For four years the Delaware and Hudson Company supported measures which it anticipated would bring Lackawanna coal into common use aboard steamboats. These were years of constant disappointment, for no inventor had designed a marine boiler which could be heated effectively by anthracite. At last the discouragement was transformed into bright optimism. The man responsible for the change was none other than the remarkable Dr. Eliphalet Nott, clergyman, educator, lecturer, and holder of more than a score of patents in anthracite stoves for home heating and cooking. In 1835 this practical improver designed tubular boilers for the ferryboat "Essex," which ran from Cortland Street Wharf to Jersey City. The boilers consisted of a large number of malleable iron tubes, each one and one-half inches in diameter and three feet in length. The tubes were installed vertically in a chamber seven feet long, seven feet high, and three and one-half feet wide. The furnace was placed alongside the chamber or boiler containing the tubes and fired by large lumps of Lackawanna coal.[30] The problem of producing a flame to create sufficient steam to propel the boat was solved with blowers which injected condensed hot air into the bottom of the firebox. The "Essex" had two of these boilers and made a trial run on March 16, 1834, up the Hudson and around New York harbor, traveling in all about forty or fifty miles. "The success was complete, and we believe satisfied all on board that the desideratum of generating steam by anthracite coal (aboard steamboats) has at length been attained," reported the *New York Journal of Commerce*.[31]

---

[27] Delaware and Hudson Minute Book, March 22, May 12, 1831.
[28] *Ibid.*, May 12, 1831; *Miners' Journal*, July 9, 1831, May 12, 1832.
[29] Delaware and Hudson Minute Book, February 6, 1834.
[30] *American Railroad Journal*, IV, March 21, 1835, p. 85.
[31] *New York Journal of Commerce*, March 18, 1835.

The Delaware and Hudson Company received the news of the "Essex" experiment with a great deal of satisfaction and reported to its stockholders that a new era would soon dawn in the Lackawanna coal trade.[32] The board of managers had good reason to rejoice. The board had followed Dr. Nott's experiments with interest. A few days prior to the "Essex" run it had drawn up a tentative agreement with H. Nott and Company, a leading stove manufacturing firm controlled by Dr. Nott. The stove company had sponsored the development of Dr. Nott's patent anthracite tubular boilers and had provided the capital necessary for their construction at Stillman's Novelty Works in New York. The agreement between the two firms stated that if Nott and Company succeeded in running a steamboat on the Hudson from New York to Albany at a speed equal to the other river boats, the Delaware and Hudson would supply the boat with 5,000 tons of coal each year for six years at four dollars per ton, or less should prices drop.[33] The price of four dollars was approximately half the average retail price of anthracite on the New York market in 1835.[34] Should the tests fail, the coal firm would have an option to purchase the patent on the boilers at a fifty per cent discount. A few days after the "Essex" experiment, the clause concerning the option in the patent was altered and H. Nott and Company agreed to sell the boilers to the Delaware and Hudson at twenty per cent discount to equip one steamboat, but retained the patent rights.[35] The success of the "Essex" had convinced Nott of the promise of his invention.

A year later, on June 23, 1836, the steamboat "Novelty," equipped with new and improved Nott anthracite tubular boilers, cast off from Chambers Street Wharf in New York. It was six o'clock in the morning. On board was a sleepy but expectant "party of gentlemen, consisting of the managers of the Delaware & Hudson . . . and others," including, of course, Dr. Nott.[36] The "Novelty" was a large steamboat, over 250 feet long. She contained twelve Nott boilers and four furnaces fed by Lackawanna anthracite, with steam blowers to stimulate combustion. The trip to Albany took twelve hours, which was considered good time. Philip Hone, first president of the Delaware and Hudson and keen observer of the men and events of his era, made the trip. He entered in his diary for that day: "Dr. Nott has suc-

---

[32] *Delaware and Hudson Company Annual Report* (1835), p. 6.
[33] Delaware and Hudson Minute Book, May 6, 1835.
[34] See Table A, Appendix.
[35] Delaware and Hudson Minute Book, May 6, 1835.
[36] Allen Nevins (ed.), *The Diary of Philip Hone* (New York, 1927), I, 214.

ceeded completely in the invention, which establishes certainly that coal will succeed wood in all our steamboats, and the Delaware and Hudson Company will hereafter be able to sell all the coal they can bring down the canal at an advanced price."[37] Hone estimated that a steamboat the size of the "'Novelty" would have consumed forty cords of pine wood at six dollars a cord, whereas the coal consumption amounted to twenty tons at a maximum price of five dollars per ton. The cash saving was apparent and Hone had good reason to be encouraged.

Anthracite coal would be of value not only to Hudson River steamboats, but to Long Island Sound steamers. Much of the deck space, heretofore cluttered with bulky wood fuel, could be cleared for cargo and passengers. Large sound steamers burned over sixty cords of wood per trip between New York and Providence and could not carry the entire fuel load. A wood sloop was picked up off Fishers Island, at the end of Long Island Sound, and the steamer took on wood for the rest of the trip while traveling at reduced speed.[38] This was dangerous and sometimes impossible in rough water, as well as expensive in money and lost time. Anthracite coal would eliminate this refueling problem.

The dawn of a new era in the Lackawanna coal trade, which the board predicted, did not appear for a few years. When it did appear, it surpassed all expectations. The Delaware and Hudson intensified the campaign for adoption of its coal in steamboats, but the conversion of boilers from wood to coal was expensive in spite of the attraction of future fuel savings through coal. It was not until 1840 that the company regarded its efforts as completely successful. Lackawanna coal was being accepted on an increasing scale by steamboats on the Hudson and along the New England coast.[39] To meet the growing demand for steamboat fuel, the Delaware and Hudson Canal was widened and deepened to permit forty-five-ton coal boats to pass. The eighteen-mile railroad was double-tracked from the mines to Honesdale at the head of the canal. By 1844 the company had increased its tonnage fifty per cent on the canal and one hundred per cent on the railroad, and still was pressed to satisfy the market at the peak of the season. In 1840 it was estimated that eastern steam-

---

[37] *Ibid.,* p. 215.

[38] R. G. Albion, *The Rise of New York Port* (New York, 1939) , p. 159.

[39] *Delaware and Hudson Company Annual Report* (1841) , 4-5; *Niles' National Register,* LVIII (July-December, 1840) , 229, 240.

boats used one hundred and fifty thousand tons of anthracite coal.[40] Much of this was supplied by the Delaware and Hudson. There is no doubt that the primary reason for the increased sales of the company was the use of Lackawanna coal by eastern steamboats.[41]

The steamboats on the Delaware River also adopted anthracite as their fuel during the forties. The Lehigh Coal and Navigation Company should be given credit for running the first anthracite steamboats. But the Lehigh Company, like the Schuylkill operators, was busy supplying domestic and industrial demand in the thirties and encouraging uses in home, factory, and blast furnace. The successful campaign for the use of anthracite in steamboats was conducted by New York and not by Philadelphia.

In 1800 it was nine days by sloop from New York to Albany. In 1850 the trip was made by anthracite steamboat in as many hours. To Americans of that time the nation stood on the threshold of new and even greater discoveries in transportation. Those who looked back into the years could say with Philip Hone, "What wondrous changes have occurred in our day and generation!"[42]

SHIPS OF THE SEA

The beginning of transoceanic steam navigation was marked by the experimental voyage of the "Savannah" in 1819, a ship as American as her name. Built in New York the year before, equipped with a steam engine constructed in Morristown, New Jersey, and boilers made in Elizabeth, she originally was destined for the run between New York and Savannah. The Panic of 1819 put an end to the coastwise career of the new ship, and her owners, the Savannah Steamship Company, decided to sell the vessel abroad. It was therefore economic necessity and little else which prompted the first crossing of the Atlantic by a ship using auxiliary steam power.[43]

The "Savannah," captained by Moses Rogers, who had commanded the steamboat "Fulton" on the Hudson, departed from her home port in Georgia on May 24, 1819. She was bound for Liverpool, but her ultimate destination was a Russian port, probably Kronstadt. Her owners hoped Czar Alexander's interest in steamboats would make

---

[40] *Niles' National Register*, LXVI (July-December, 1844), 281.

[41] *Fisher's National Magazine*, I, August 1845, p. 220.

[42] Nevins, *Diary of Philip Hone*, II, 905.

[43] Tyler, *Steam Conquers the Atlantic*, pp. 7-11.

him receptive to the demonstrated prowess of the "Savannah." For her crossing the ninety-eight-foot vessel took aboard fifteen hundred bushels of coal, or about fifty-four tons, and twenty-five cords of wood.[44] The coal was not from Pennsylvania, but was either English or Virginia bituminous.

The voyage took twenty-nine days and eleven hours from Savannah to Liverpool, including time lost at Kinsale, Ireland, where more coal was taken on to replenish exhausted supplies. The ship was to burn approximately ten tons of coal per day. Steam power was employed only a total of three and one-half days of the passage.[45] The fact that the ship exhausted fifty-four tons of coal shortly before reaching Liverpool, after using her engines only eighty or ninety hours, employing sail most of the trip, indicates the estimated consumption of ten tons per twenty-four hours was entirely too low. It proved to be too low because of the inefficiency of engine and boilers and the consequent waste of fuel in building steam pressure. Also, steam was used at varied intervals during the crossing. This consumed more fuel, since larger amounts of wood and coal had to be used to kindle the fire and gain steam pressure. When the engine was shut down, insufficient combustion of the coals on the fire bed meant additional waste. To cross the Atlantic using steam power necessitated drastic technical improvements in coal-burning boilers and steam engines. Certainly, no ship could hope to carry enough wood for the trip, and the experience of the "Savannah" left little hope that a small ship could carry enough coal. At the best, steam power for transoceanic voyages was to remain a minor auxiliary to sail until further progress was made in steam marine engineering.

The "Savannah" found no purchasers in England, Sweden, or Russia. Discouraged, Captain Rogers sailed back to the United States. Coal was expensive and wind was free, but the ship hit the stubborn westerlies and took forty days from Norway to Georgia. Using sail all the way across the Atlantic, Rogers gave no orders for the use of steam power until he entered the mouth of the Savannah River. Rejected by the federal government and by European countries, the ship met with a series of calamities and finally was sold at auction. Her engine was taken out and used as a stationary steam engine by a New York factory. The "Savannah" became nothing more than a coastwise sailing ship.[46]

---

[44] *Ibid.,* p. 11.

[45] *Ibid.,* p. 12.

[46] *Ibid.,* p. 13.

After this initial endeavor, Americans did not attempt to brave the Atlantic with steam-powered vessels for more than twenty years. Instead, they turned their attentions to steamboats on inland waters. The British, blessed with capital and coal, led the way in Atlantic steam navigation, but even their endeavors did not mature until the late thirties. In 1838 the "Sirius" and the "Great Western" plodded their way across the sea to New York and were given gala receptions by the citizens of the city. The "Sirius" burned over 450 tons of British bituminous on her maiden voyage. The "Great Western" burned Welsh bituminous. On later voyages she experimented with Pennsylvania anthracite, though not too successfully, for she soon reverted to bituminous coal. The "Liverpool," another early British steamship, burned samples of American anthracite.[47] Though the press reported satisfaction with anthracite, the common practice of the British steamship was to use bituminous coal. Many complained of "inferior" American coals and, despite the duty, established fuel depots in United States ports and stocked them with British coal. It was not until Maryland Cumberland coal, an excellent bituminous steam fuel, became available in quantity during the fifties that British steamships, especially the Cunard Line, took advantage of American coal on a large scale.[48]

The United States clung to sail. For a decade after the initial successes of the British there were no American experiments in ocean steam navigation. Reluctance to enter this field had nothing to do with the amount of available coal, for the market could have been supplied by the anthracite producers. The United States lacked both capital and public confidence to encourage the immediate founding of steamship lines to compete with Britain. Early but futile efforts of some farsighted businessmen, including Nicholas Biddle, met with cool reception. The energies of the Delaware and Hudson were absorbed introducing anthracite aboard river and sound steamboats. The company paid little attention to the possibilities of a new market until the fifties when ocean steam navigation had reached a mature level.[49]

The coal interests of the Lehigh and Schuylkill regions, however, showed considerable interest in the potential market for anthracite

---

[47] Hazard, *U. S. Register,* I, July 10, 1839, p. 34.

[48] *DeBow's Review,* new series, I (1853), p. 476; *Thirty-fourth Annual Report of the President and Directors to the Stockholders of the Baltimore and Ohio Railroad Company* (1860), pp. 23-24. Referred to hereafter as *Annual Report, B&O.*

[49] *Delaware and Hudson Company Annual Report* (1852), p. 4.

aboard ocean steamships.[50] Word drifted to Philadelphia that the "Great Western" considered anthracite superior steamship fuel. The Philadelphia *North American* jumped to the conclusion that Pennsylvania would soon become the source of all fuel energy for steamers standing out of American ports. "The giant Pennsylvania . . . her bowels . . . filled with . . . coal," was favored by God as "the repository of untold wealth and blessings."[51]

On March 11, 1841, the British steamship "President" left New York bound for England. She was never seen again. The mysterious disappearance of the vessel and the 110 persons aboard provoked endless speculation in the press. Benjamin Bannan's *Miners' Journal* seized upon the incident to offer its opinion on the fate of the lost ship. The *Journal* recalled that the "President" carried bituminous coal in her bunkers. Bituminous coal, said the paper, was susceptible to spontaneous combustion. The ship "probably" took fire and went down somewhere in mid-ocean. This was not all the *Journal* had to say. It went on to relate that the "British Queen," the "Great Western," and two of the Boston steamers had had fires in their bunkers, but the facts were withheld from the public.[52] These serious charges were caught up immediately by other newspapers. Some felt that spontaneous combustion did or could occur, but most of the papers saw through the bias of the anthracite trade organ. The *New York Herald* and the *Philadelphia Ledger* were particularly vehement in their criticism of the *Miners' Journal*. This attitude "might be expected," retorted Bannan, "from a prostituted and venal press," but the honor of Pottsville was offended when the *Boston Transcript* called the thriving center of the Schuylkill coal trade "a back-country village."[53] Benjamin Bannan quoted a dozen instances of spontaneous combustion in vessels carrying bituminous coal. Not only had steamships experienced these unwelcome disasters, but sailing ships carrying soft coal as ballast found the fuel a fire hazard. Scientific opinion was brought to bear in support of the *Journal's* arguments, and the paper even went so far as to demand laws prohibiting the use of bituminous coal in steamships using American ports.[54] The Philadelphia

---

[50] *Report of the Board of Managers of the Lehigh Coal and Navigation Company,* January 14, 1839, p. 32; *Miners' Journal,* October 23, 1841.

[51] *North American.* Reprinted in Hazard, *U. S. Register,* I, September 25, 1839, p. 215.

[52] *Miners' Journal,* July 24, 1841.

[53] *Ibid.,* August 7, 1841.

[54] *Ibid.*

North American, a high-tariff, pro-anthracite paper, supported some of the charges against bituminous coal and cited about twenty occurrences of spontaneous combustion in bituminous coal piles in yard bunkers and storage areas.[55] The Philadelphia Gas Works was annoyed with this problem for many years.

The public became genuinely concerned over these tales. By October, 1841, American passenger travel aboard British bituminous steamships was very light.[56] For a time, fear of fire at sea drove many back to sailing vessels. Over the years ideas were advanced to solve the problem of spontaneous combustion of bituminous coal piles aboard ship, but no concrete solution was reached.[57] As long as steamers carried bituminous coal in enclosed spaces the danger was always present. The fears in the public mind were gradually dispelled as no major disasters were traced to the menace of bituminous fuel. The anthracite press abandoned the issue in 1842, but when Cumberland bituminous coal from Maryland mines gained ascendancy in steam navigation, the *Journal* again took up the cry. Ten years after its first attacks the paper informed its readers that not all bituminous coal was liable to spontaneous combustion, but Cumberland was, and it would be better to "ship aboard a powder magazine" than a vessel carrying most brands of bituminous coal.[58]

In the fall of 1841 the steamship "Clarion" was launched in New York. She was equipped with anthracite boilers and the Ericsson propeller. The *New York Herald* atoned for its earlier attitude by noting a British Admiralty report on fire by spontaneous combustion of bituminous coal on board East India steamers. The New York paper then praised anthracite as an excellent fuel for ocean steam navigation. "It has long been urged by grave authorities, that nature has imposed an effectual barrier to prevent the United States from competing with Great Britain in steam navigation, owing to the scarcity and inferior quality of our bituminous coals." The trial run of the "Clarion" illustrated the "absurdity of this opinion," said the *Herald.* Anthracite would become the steamship fuel of the future! Pennsylvania anthracite could produce steam with efficient, economical results and with an absence of smoke. Steamers burning bituminous coal could be tracked for seventy miles at sea long before their hulls were

---

[55] *Ibid.,* September 11, 1841.
[56] *Ibid.,* October 2, December 11, 1841.
[57] *Journals of the Franklin Institute,* XXXIV (July-December, 1842), 420-421; LIII (January-June, 1852), 419; *The Mining Magazine,* III (1854), 567.
[58] *Miners' Journal,* March 29, April 5, 1851.

visible because of the black coal smoke trailing along the horizon. In time of war this would make the bituminous-burning warship inferior to the anthracite vessel. The newspaper concluded the article by expressing regret that the two new United States war steamers, the "Mississippi" and the "Missouri," had been designed to burn only bituminous coal.[59] The regret was shared by anthracite operators in the Schuylkill region, and since there was a lack of American bituminous steam coal on the East Coast, it was thought the two warships would have to depend on foreign bituminous.[60]

Anthracite interests looked to the future and hoped ocean steamship lines would realize the numerous advantages of hard coal as a safer, cleaner, more economical fuel. More freight and less coal could be carried by a ship burning anthracite. Passenger packets would be able to eliminate sparks and odors which plagued travelers aboard ships using soft coal. War steamers would have a greater cruising range. The day would come when all ocean steamers would burn Pennsylvania anthracite. "What a vast field for its consumption," chorused the anthracite operators.[61] But in the forties only a few ocean steamers were built to use anthracite, and it was not until the succeeding decade that American ocean steamers used Pennsylvania hard coal in large enough quantities to be considered an important market.

With the dawn of British transoceanic steam navigation in 1838, a delegation of four naval officers arrived in England. They had been sent by the autocrat Nicholas I, czar of Russia. The purpose of the mission was to learn everything possible about British steam navigation, shipbuilding, and railroads. The Russian officers were not welcome.[62] The British had no desire to share technical secrets with the one country which constantly threatened to overturn the balance of power in the eastern Mediterranean by conquest or partition of the tottering Ottoman Empire. Rebuffed in England, the Czar then ordered the officers to the United States. Here they were greeted cordially by government officials and private citizens. Led by Captain von Schantz, the mission remained in the United States for two years,

---

[59] Reprinted in Hazard, *U. S. Register,* V, October 20, 1841, p. 253.

[60] *Miners' Journal,* October 30, 1841.

[61] *Ibid.,* October 23, 1841.

[62] *New York Herald.* Reprinted in Hazard, *U.S. Register,* V, October 6, 1841, p. 210; for a more detailed account of the Imperial Russian government's mission to the United States to obtain technical assistance and information of a military nature, see Frederick M. Binder, "American Shipbuilding and Russian Naval Power, 1837-1841," *Military Affairs,* XXI, No. 2, Summer, 1957.

touring the entire country and Texas. The Russians visited every naval station, dry dock, shipyard, and locomotive works they could find. They made over three hundred detailed drawings of steamboats, steamships, stationary steam engines, dry docks, and locomotives, which the Americans proudly exhibited to them. Evidently impressed by American technology in steam engineering and in the use of anthracite coal to produce steam, Von Schantz conferred with George L. Schuyler, a Jersey City shipbuilder, on the construction of an anthracite steam frigate. The Russians drew up the basic plans for the ship, but as the *New York Courier* pointed out, the 210-foot "Kamchatka" was Russian above the water line, but American below. The *Courier* claimed there was not a scrap of foreign metal in her. The 600-horsepower engine was built by H. R. Dunham and Company and equipped with anthracite-burning copper boilers without blowers.[63] On the trial run outside Sandy Hook she steamed "twelve miles per hour," burned clean, efficient anthracite coal exclusively, and was calculated to consume twenty-seven tons per day. Her bunkers were large enough to carry a month's supply.[64] The elimination of blowers cut fuel consumption to a minimum, although it took longer for the ship to get up steam.

When the Czar's luxurious, armed frigate steamed out of New York harbor for Kronstadt, George L. Schuyler was aboard. At Southampton he sent a "log letter" to the *New York Courier* describing the voyage. The "Kamchatka" had hit heavy seas and winds of hurricane force, but had come through the storm "almost unscathed." The ship put into England to take on additional coal "as we wished to carry some of our Anthracite to Russia," wrote Schuyler.[65] The Russians would be able to supply their new ship with anthracite coal from the lower Don Valley,[66] but Schuyler no doubt wanted to show the Russian government the "superior Pennsylvania brand." Schuyler's high opinion of Pennsylvania anthracite coal for steamships was bolstered by the successful voyage across the Atlantic.[67]

The report on the "Kamchatka" encouraged anthracite operators, but the United States did not venture into the field of trans-Atlantic

---

[63] Reprinted in Hazard, *U. S. Register,* V, October 13, 1841, p. 236.

[64] *New York Herald.* Reprinted in Hazard, *U. S. Register,* V, October 6, 1841, p. 209.

[65] *New York Courier.* Reprinted in Hazard's *U. S. Register,* V, November 17, 1841, p. 319.

[66] Taylor, *Statistics of Coal,* p. 623.

[67] *American Railroad Journal,* I, May, 1843, p. 148.

steam navigation for a few more years. The "Massachusetts," the first American steam packet since the "Savannah" to make the voyage from the United States to England and return, did not stand out of New York harbor until 1845. The voyage to England took seventeen and one-half days, most of which was made under sail despite the assertion of her captain a few years later that eleven days were made under steam. The ship bunkers simply did not carry enough coal for her engine to be used that length of time. The "Massachusetts" was not a profitable venture and competition with the British forced her owners to sell her to the United States Army. Used as a troop transport during the Mexican War, she was transferred to the Navy Department with the establishment of peace in 1848.[68] A year before the maiden voyage of the "Massachusetts," the American steam vessel, "Midas," rounded Cape Horn bound for Hong Kong. Her boilers were in such poor condition when she arrived off the China coast that she made the return voyage under canvas.[69] Two steamship lines, one to Bremen and the other to Le Havre, were given subsidies by Congress in 1847 and 1848. The lines, each boasting two ships, were a disappointment to all, including their respective founders, Edward Mills and Mortimer Livingston. They did not possess the speed to "drive the Cunarders off the ocean." Cunard's ships running from England to Boston and then, later, to New York, were the steam-queens of the sea. It was particularly galling to Pennsylvania anthracite interests that the coal burned by the Cunarders was not from their mines. Indeed, in the first few years of the life of the Cunard company, its vessels did not burn American coal of any kind. Instead of using American soft coals, sailing ships brought supplies of soft coal from Liverpool to the Cunard docks at Boston and to the newer piers across the Hudson at Jersey City.[70]

These early beginnings in American ocean steam navigation did little to enhance the Pennsylvania coal market. It was not until the appearance of the ill-fated Collins Line in 1850 that Pennsylvania anthracite gained a steady customer in trans-Atlantic shipping. Collins first experimented with Dauphin semibituminous coal from the Susquehanna Valley and found it superior to Cumberland coal from Maryland.[71] The line's four steamers, the "Atlantic," "Pacific,"

---

[69] *Scientific American*, X, October 28, 1854, p. 51; *Journal of the Franklin Institute*, LVI (January-June, 1853), 57.

[69] *Scientific American*, X, October 28, 1854, p. 51.

[70] Albion, *The Rise of New York Port*, pp. 324-326.

[71] Taylor, *Statistics of Coal*, pp. 80-81.

"Arctic," and "Baltic," soon turned to the readily available Pennsylvania anthracite and found it to be more satisfactory. Collins steamers burned Pennsylvania anthracite on their voyages from New York to Liverpool, and Welsh coal on their return trips.[72] B. F. Isherwood, chief engineer, United States Navy, studied the steam log of the "Arctic" and stated that the Welsh coal burned on voyages was "the Welch [*sic*] coal of similar chemical composition," meaning, of course, that it was Welsh anthracite.[73] In the fifties the Cunard steamers burned Welsh bituminous on their voyages from England to the United States, but had abandoned the expensive procedure of sending coal supplies to this country by sailing sloop for their return trips. Instead, Cunard began using Cumberland bituminous which, by that time, was shipped in large quantities from Baltimore to the northern seaports.[74]

It was this Maryland bituminous coal which became the competitive fuel of anthracite in trans-Atlantic steam navigation. Cunard ships, whose boilers were not constructed to burn anthracite, considered it to be superior to Virginia or Nova Scotia soft coal and equal to Welsh bituminous. Those who supported the Cumberland interests claimed that it was superior to anthracite in producing steam. The Collins steamer "Pacific" had made several fast runs with Cumberland coal and for a short time the company leaned toward the Maryland product. When B. F. Isherwood carefully analyzed the steam log of the "Pacific," he reported that the speed attributed to the use of Cumberland was due to the excellent weather conditions during the crossing and not to the superior qualities of the fuel. Isherwood advocated the continued use of anthracite in the Collins steamers.[75] The line took his advice until its tragic end through shipwreck and financial failure, the latter due to the termination of the company's subsidies by Congress in 1858. From that time on trans-Atlantic steamship traffic was in the hands of European concerns whose ships' boilers were not designed to burn anthracite. These ships did use Cumberland coal, however, and some Pennsylvania soft coals which found their way to the seaboard in the late fifties.[76]

In spite of the victory of Europe in trans-Atlantic steam navigation, more than half the steamships plying the United States coastal trade in

---

[72] *Ibid.,* p. 85; *Miners' Journal,* February 12, 1853.

[73] *Journal of the Franklin Institute,* LVI, 41.

[74] Taylor, *Statistics of Coal,* p. 85; *Miners' Journal,* February 12, 1853.

[75] *Journal of the Franklin Institute,* LVI, 400-401.

[76] *Thirty-fourth Annual Report, B&O* (1860), pp. 23-24.

the fifties used Pennsylvania anthracite coal in whole or in part, while steamers out of New Orleans usually burned Pittsburgh bituminous. Anthracite was shipped to Havana and to Nicaragua to supply American steamships on the Caribbean run.[77] Anthracite was even sent around the Horn or across the Isthmus to California to feed the boilers of steamers on the Pacific coast.[78] By the mid-fifties the coastal steamers, Caribbean ships, and the four Collins vessels used an estimated quarter of a million tons of Pennsylvania anthracite annually.[79]

The use of steam and coal by ocean shipping placed sail in a subordinate position. In time the merchant steamer would drive even the sleek "clipper" from the seas. But the story of the development of the American steam navy took no such glorious course and lagged behind private shipping interests in the utilization of steam propulsion.

Advocates of steam-powered war vessels were few in the service of the United States Navy. Most of the older officers resisted steam navigation in the late thirties. This trend was to continue for a decade beyond the Civil War. Shades of the sea victories in the War of 1812 haunted memory and tradition and blurred the vision of the men who loved the sailing ships. Not only naval officers, but high government officials showed disdain for steam navigation. President Van Buren evidenced little interest and some hostility to the development of naval steamships. The President's attitude was magnified in the stand taken against steam by the Secretary of the Navy, James Kirke Paulding, who looked upon steamships as "sea monsters." After the launching of the navy's first steam vessel, the "Fulton," in 1837, there would have been little progress had not some public concern shown itself in Congress over the construction of steam navies by England and France. In 1839 Congress authorized the construction of three more steam vessels.[80]

The former captain of the "Fulton" returned from Europe that same year. Captain Matthew C. Perry, remembered for his famous expedition to Yedo Bay in the fifties, was one of the foremost pioneers of steam navigation in the United States Navy. His European tour

---

[77] *DeBow's Review*, I (1853), p. 476.

[78] *Cannelton, Perry County, Indiana*, p. 69.

[79] *DeBow's Review*, I (1853), p. 476. The exact estimate was 200,000 tons for 46 steamers in 1853; Captain Charles Wilkes, USN, *Report on the Examination of the Deep River District, North Carolina* (1858), p. 26.

[80] Harold and Margaret Sprout, *The Rise of American Naval Power 1776-1918* (Princeton, 1946), pp. 112-114.

had been devoted to the study of steam engineering in the navies of England and France. On his return, Perry's technical knowledge and experience largely contributed to the design and construction of two of the three steamships completed in 1842.[81] These were the "Mississippi" and "Missouri," whose boilers were equipped to burn bituminous coal—a fact deplored and lamented by the *Miners' Journal* and the *New York Herald*.[82] The two frigates, each 229 feet long, possessed small auxiliary steam engines. It was obvious that the ships were meant primarily for sail rather than for steam. The extension of the Baltimore and Ohio Railroad to Cumberland coincided with the launching of the vessels in 1842 and made the Cumberland coal of western Maryland available to the East Coast. Seventeen thousand bushels were purchased by the Navy Department for the two steamers.[83]

At this time the Navy Department also sent requests to anthracite operators for samples of coal to be sent to the Navy Yard at Washington, D.C., and to other naval establishments on the Atlantic and Gulf coasts and on the Lakes. The sample shipments, of not less than two tons each, were to be forwarded at the expense of coal operators with a statement of the coal's origin and time of extraction.[84] The anthracite, no doubt, was to be tested in stationary engines in the Navy Yard as it could not be burned by the new steamers. But the Schuylkill interests were delighted, and the *Miners' Journal* voiced confidence in the federal government's drive on fuel economy by proclaiming "now that the Rubicon is passed they can have every proof of the efficiency and superiority of Anthracite over any other."[85]

The third ship to be built under the Act of 1839 was the first screw-driven warship in any navy in the world. Behind the construction of this vessel were three men: Captain Robert F. Stockton, of Princeton, New Jersey; John Ericsson, inventor of the submerged screw propeller; and Abel P. Upshur, Secretary of the Navy and later Secretary of State. Upshur was a staunch supporter of the steam navy idea.[86] Built at the Philadelphia Navy Yard and launched in 1844, the ship, named the "Princeton," was to come into national focus and to be remembered by students of American history for reasons totally

---

[81] *Ibid.*, p. 114.

[82] Hazard, *U. S. Register*, V, October 20, 1841, p. 253; *Miners' Journal*, October 30, 1841.

[83] *Niles' National Register*, LXII (1842), p. 112.

[84] *Miners' Journal*, April 23, 1842; *The Pennsylvanian*, May 21, 1842.

[85] *Ibid.*

[86] Sprout, *American Naval Power*, p. 125.

different from her revolutionary design. The tragic cruise of the "Princeton" up the Potomac and the explosion of her new cannon, the "Peacemaker," ended the lives of six persons, one of whom was the Secretary of the Navy, Thomas W. Gilmer, and another, the Secretary of State, Abel P. Upshur. It was an ironic twist of events when the shattered fragments of an exploding gun aboard the newest steam vessel of the navy snuffed out the life of Upshur. He had been in the front ranks of the few men who campaigned for technical progress and modernization in the navy of the United States. The glory of the "Princeton" was dimmed in the public eye, but she was recognized as the leader in naval engineering by England and France. Not only was she equipped with the screw propeller, but her boilers had been built to burn Pennsylvania anthracite. In 1845 she proved the excellence of Pennsylvania hard coal in speed tests and, during the Mexican War, used anthracite successfully when operating in the blockade of Veracruz.[87]

While Upshur was still Secretary of the Navy, complaints were received by his department concerning the coals procured to service the new bituminous-burning vessels. The navy commissioned Professor Walter R. Johnson, a prominent engineer and fuel analyst, to determine the best coal for naval use. Johnson issued his famous controversial report to the Navy Department on American coals in 1844, a few months after the deaths of Upshur and Gilmer. In his research, Johnson found a twofold problem. He had to select the best coal from the standpoints of both price and efficiency. Ability to raise steam, durability of grate bars, and a dozen other properties were analyzed by the scientist and his staff in a specially constructed laboratory. Johnson experimented with forty-one samples of coal, nine of which were Pennsylvania anthracite. He did not limit his tests to coals found in the United States, but also analyzed samples of foreign coals. One of the many ways in which these experiments differed from previous ones was in the amount of coal used. Johnson utilized several hundred pounds of each kind in every test. Marcus Bull, many years before, in conducting his experiments, had used only a pound or two of each type of coal.[88]

Johnson himself admitted that his experiments were by no means conclusive, but to his satisfaction he had scientifically proven that the most efficient coal for naval steam vessels was Maryland Cumberland

---

[87] *Journal of the Franklin Institute,* LVI, 43-50.

[88] *Senate Document,* No. 386, 28th Congress, 1st Session, pp. i-xi.

from the Atherson and Templana mines. The report was voluminous. When published it consumed over six hundred pages. It was submitted to the Senate by James Y. Mason, Secretary of the Navy. Evidently the Senate considered the report significant enough to order 10,000 copies printed and distributed throughout the United States.[89]

As a result of Johnson's findings the navy continued to burn Cumberland coal. Even the "Princeton" was eventually converted to burn this fuel. The gradual growth of the steam navy in the following years to sixteen vessels by 1853, and the successful use of anthracite in coastwise and transoceanic steam navigation, reopened the fuel controversy in naval circles. Congress instructed the Secretary of the Navy to obtain a report from the Engineer in Chief, Charles B. Stuart, on the comparison of anthracite with bituminous in naval steamers. Stuart submitted his report to the secretary in February, 1852. In the spring of that year he elaborated some of his findings in a letter addressed to the Chairman of the Committee on Naval Affairs.

Professor Johnson's experiments of the preceding decade were literally torn to ribbons by Stuart's practical findings. The coal used in the navy tests was Cumberland bituminous and white-ash Schuylkill Valley anthracite. Emphasizing throughout his letter that the only true test for comparative value of coal was to take the fuels as they were delivered to the ships, stored in the bunkers, and brought out for use, Stuart criticized Johnson's limited experiments. Johnson had used, at the most, no more than half a ton in each laboratory test and in a boiler which was not used by naval steam vessels. The navy used hundreds of tons in marine boilers under actual steaming conditions over a period of time. Stuart found Cumberland coal to be more expensive than anthracite, costing about $1.50 more per ton at the New York Navy Yard. Since the former coal was friable, a portion of it always crumbled into unusable powder in the process of loading and handling, thus increasing the real cost of the fuel. What's more, it was liable to spontaneous combustion, a hazard of great concern aboard naval craft. In comparison, not only was the initial cost of anthracite less, but the hard coal was easier to handle and not liable to spontaneous combustion. Under cruising conditions it had been proven to be one-third more effective than Cumberland coal in getting up steam rapidly and sustaining it.

Stuart also pointed out that a ship could steam two-thirds longer with its bunkers stocked with anthracite. Because of its smaller bulk

---

[89] *Ibid.*, pp. 599-600.

more of this coal could be taken aboard. This was of great importance to the navy as it meant a longer time could be spent at sea
without the necessity of refueling. Smoke from the stacks of
bituminous-burning naval vessels could be tracked for miles at sea
before their hulls were visible, and their positions easily determined,
by enemy ships. Anthracite not only threw out less smoke, but fewer
sparks, minimizing the danger of fire. The intense heat of anthracite
in marine steaming made copper boilers impractical; thus, in closing,
Stuart recommended iron boilers whenever possible. The chief engineer's report seemed confirmed by additional observations made by
another naval technician, B. F. Isherwood, who had studied the use
of Pennsylvania and Welsh anthracite on a cruise from the Battery
to Liverpool, as well as by earlier British experiments which had reported anthracite's superiority over bituminous coal in naval vessels.[90]

The controversy was by no means settled, but the navy leaned toward anthracite fuel from Schuylkill County from 1852 through the
Civil War. It should be pointed out, however, that prior to the Civil
War steam was regarded merely as an auxiliary power to sail, to be
used in battle maneuvers, in calms, or in entering or leaving port.
The war demonstrated the many practical advantages of steam-
powered vessels during the tedious blockading operations by the
Union. *The Records of the Union and Confederate Navies* is filled
with letters, orders, and dispatches regarding coal supplies for Northern ships and the difficulties experienced in securing coal by the
proud, destructive commerce raiders of the Confederacy.[91]

Most of the coal used by the Union Navy came from the Pennsylvania mines. The majority of the Pennsylvania tonnage was anthracite brought by sailing sloop from New York or Philadelphia to
coaling stations or lighters along the eastern coast, Cuba, and the
Gulf Coast of Florida.[92] Henry C. Carey, in 1867, bemoaning the
financial losses of anthracite coal operators, no doubt exaggerated
anthracite's part in the war, but his words are worth quoting: "But
for Pennsylvania anthracite . . . the cause of the North would this

---

[90] *Senate Executive Document,* No. 74, 32nd Congress, 1st Session, pp. 1-14; *Journal
of the Franklin Institute,* LIII, 418-419; LIV (July-December, 1852), 217-228; LV,
40-41.

[91] For examples see *Official Records of the Union and Confederate Navies in the
War of the Rebellion* (Washington, 1894), Series I, Vol. I, pp. 156, 477, 479, 560-561,
647; Series I, Vol. II, pp. 82, 257-259, 353-354, 605, 627.

[92] *Ibid.*

day be 'the lost cause.' "[93] "Pennsylvania," continued Carey, "alone in the possession of anthracite, . . . furnished nearly all the motive power that maintained the blockade."[94]

Despite strong dependence on steam during the Civil War, the navy returned to canvas with the peace. By 1870, under the navy regulations, steam power was to be used only when absolutely necessary. Captains were required to enter in their logbooks in red ink the reasons for getting up steam and starting engines.[95] Admiral Porter, staunch defender of fighting ships of wood and canvas, was bitterly opposed to Isherwood, chief of steam engineering.[96] The Admiral even suggested that the regulation requiring less steam and more sail could be enforced by charging the cost of the coal consumed against the commanding officer's pay.[97]

Thus, a decade after the end of our period, one finds Pennsylvania coal in the shadow of sentiment and sail, neither the fuel nor the steam it produced fully accepted by the United States Navy.

---

[93] *Works of Henry C. Carey* (Philadelphia, 1867), XXIX, "Letter Sixth," H. C. Carey to Senator Henry Wilson, September 10, 1867, pp. 28-29.

[94] *Ibid.*, "Letter Seventh," pp. 32-33.

[95] Sprout, *American Naval Power*, p. 167.

[96] J. T. Morse, Jr. (ed.), *The Diary of Gideon Welles* (Boston, 1910), III, 283.

[97] Sprout, *American Naval Power*, p. 168.

CHAPTER VI

The Coal-burning
Locomotive in America

PIONEER EXPERIMENTS WITH LOCOMOTIVE FUEL

THE history of the coal-burning locomotive in the United States begins in 1830 with Peter Cooper's "Tom Thumb," the first American-built locomotive to run in this country. The "Tom Thumb" did not burn wood as is popularly believed today, but used Pennsylvania anthracite coal.[1] *The Eighth Annual Report of the Baltimore and Ohio Railroad Company* for the year 1834 contains the following lines: "In the year 1830, Peter Cooper, Esq., illustrated by an experiment with a small working locomotive engine, with a tubular boiler, the practicability of using Anthracite as fuel."[2] Combustion was aided by an artificial draft in the firebox. The draft was created by a fan driven by a belt passed around a wooden drum which, in turn, was attached to one of the road-wheels.[3]

The Baltimore and Ohio Railroad Company had been organized in 1827 by a number of progressive Baltimore citizens. It was believed that the construction of a steam railroad "from Baltimore to some eligible point on the Ohio River" would tap the western trade and make the city second only to New Orleans as the great outlet of the West. Strickland's report to the Pennsylvania Society for Internal Improvements was referred to at length and the success of railroads and coal-burning locomotives in England was used as a persuasive argument for founding an American railroad.[4]

---

[1] "The first practical American steam locomotive was the Tom Thumb, a high-stacked wood-burner built in 1829 . . . railroad men used wood-burning engines until after the Civil War when mining became more economical and locomotives could afford to use coal." *New York Times*, February 14, 1954.

[2] *Eighth Annual Report, B&O* (1834), p. 30.

[3] W. H. Brown, *The History of the First Locomotives in America* (New York, 1871), p. 112.

[4] *Proceedings of Sundry Citizens of Baltimore Convened for . . . Improving the Intercourse Between that City and the Western States* (Baltimore, 1827), pp. 3-5, 9, 31.

Thus, from the inception of the company, it was decided to follow the English example and use mineral fuel in the locomotive engines. A year later when the engineers made a survey report to the directors, they noted, with satisfaction, that the proposed route would cut through country where there was easily accessible bituminous coal. This was the Cumberland area of western Maryland. At that time Cumberland coal was brought down the Potomac and used by the armory at Harpers Ferry and burned in the homes of Hagerstown and Williamsport.[5] The railroad company, which had yet to lay a track, already was contemplating Cumberland coal as the source of fuel for their locomotives and as profitable freight to supply the eastern seaboard. It was to be more than a decade before these things were realized.

In 1828 the Delaware and Hudson Company had sent its agent, Horatio Allen, to Great Britain to purchase several locomotives to be used on a short railroad line the company had built from its mines at Carbondale to Honesdale at the head of the canal.[6] The next year one of these locomotives, the "Stourbridge Lion," had the distinction of being the first steam locomotive to operate in the United States, but was "retired" after two short runs of three miles. The Delaware and Hudson then abandoned, for the time, its project for a steam railroad. The "Stourbridge Lion" had proved to be too heavy for the iron plates on the wooden rails and tore them off as it rounded the curves of the road.[7] It is not known if the "Stourbridge Lion" burned coal or wood.

Contrary to the Delaware and Hudson, the Baltimore and Ohio turned its back on the British and encouraged American inventive talent. In 1830 it had experienced some success with the "Tom Thumb" and shortly thereafter issued a public call to "American genius" to perfect a large, practical coal-burning locomotive.[8] In 1831 three American-built locomotives were produced, but the most creditable performance was given by the "York," a three and one-half-ton engine constructed by Phineas Davis. This locomotive could haul fifteen tons at fifteen miles per hour on a level track. Also, it could attain the phenomenal speed of thirty miles per hour without a load on the straight sections of the road. "The fuel used is anthracite coal, which has been found to answer the purpose well," read the chief

[5] *Second Annual Report, B&O* (1828), pp. 7-8, 77.
[6] Delaware and Hudson Minute Book, January 16, 1828.
[7] *Ibid.,* August 13, 1829.
[8] *Fifth Annual Report, B&O* (1831), p. 23.

engineer's report.[9] But the locomotive was too light to use on ascents, and the company requested the inventor to build a heavier one.[10] The firm of Davis and Gartner, of York, Pennsylvania, of which Phineas Davis was the senior partner, proved equal to the task. The following year Davis experimented with the six and one-half-ton "Atlantic" on the Baltimore and Ohio tracks.[11] The "Atlantic" averaged twelve miles per hour as it rolled over the forty miles of newly laid track from Baltimore to Parrs Ridge. The round trip of eighty miles was accomplished on one ton of anthracite coal, which the company had purchased for eight dollars. The chief engineer was well pleased with the performance of the new engine and with the anthracite coal, which he felt was more economical than coke, and equal to it in the fact that it produced little or no smoke.[12] The board of directors also was pleased with the experiment and with the subsequent performance of the locomotive.

Davis and Gartner had bought the Peter Cooper patent to the vertical boiler. This model was important in the use of anthracite for fuel "where the intense heat of anthracite coal renders every protection necessary which can be afforded to the bottom of the boiler and to the tubes." In a vertical boiler, expanding steam pressure continually pushed more water toward the fire at the bottom of the tank and protected the tubes. The "Atlantic" ran thirteen hundred miles during the first year and did not burn out a single tube. The locomotive continued to use "anthracite coal with great facility and powerful effect." With a finality which seemed to indicate the end of fuel experimentation, the engineer reported, "Nothing more is wanted to demonstrate the entire practicability and utility of this fuel for the locomotive engine."[13] By 1833 the Baltimore and Ohio was paying seven dollars per ton for Pennsylvania anthracite which came by way of the treacherous Susquehanna River, or from Philadelphia via the Chesapeake and Delaware Canal. From the very beginning the company had ignored cheap wood fuel and had not attempted to use Virginia or foreign bituminous coal. Anthracite was felt to be the only safe, efficient combustible for producing steam in locomotive engines.[14]

---

[9] *Ibid.*

[10] *Ibid.*, p. 24.

[11] *Sixth Annual Report, B&O* (1832), p. 9.

[12] *Ibid.*, pp. 49, 109.

[13] *Seventh Annual Report, B&O* (1833), pp. 11-12, 34.

[14] *Ibid.*, pp. 104, 116, 178.

Davis and Gartner had built three more anthracite locomotives for the Baltimore and Ohio by 1834.[15] One of these, the "Arabian," proved to be the death trap for its inventor, Phineas Davis, the first man to construct a practical locomotive in which anthracite fuel was "successfully" employed. Davis was killed in a railroad accident in September, 1835, on a trial run to Washington while riding the tender of his new engine.[16] With the death of the inventor, the firm of Davis and Gartner was dissolved.[17] Davis also had been superintendent of the Baltimore and Ohio locomotive shops at Mount Clare. These shops then were leased by Gillingham and Winans, who continued to build anthracite locomotives for the railroad.[18] Ross Winans was to become one of the leading locomotive manufacturers of his age, challenging even the great Baldwin of Philadelphia and certainly surpassing him in the construction of coal-burning engines.

It was in 1837 that the Baltimore and Ohio, pioneer in the use of anthracite fuel for locomotives, began to consider the use of bituminous coal. The railroad was to extend beyond Harpers Ferry into the Cumberland coal country as the company sought its ultimate objective, the Ohio River. Uncertain of the fitness of bituminous coal as locomotive fuel on the harsh grades and short curves of American railroads, the Baltimore and Ohio conducted a fuel survey of ten United States rail lines. The only road of consequence which was using coal other than the Baltimore and Ohio was the New York and Paterson, which burned Beaver Meadow anthracite.[19] There was another road which began using hard coal in 1837. This was a short spur line from the anthracite region near Mauch Chunk known as the Beaver Meadow Railroad. Either the Baltimore and Ohio was ignorant of its existence or considered the road of so little consequence that it did not bother to include it in the survey. The Beaver Meadow line was only fourteen miles long. Its first coal-burning locomotive was built by Garrett, Eastwick and Company in 1837.[20] The other roads in the survey, including the Boston and Providence, the Boston and Worcester, and the Camden and Amboy, burned pine wood under horizontal boilers. By this date the Baltimore

---

[15] *Eighth Annual Report, B&O* (1834), pp. 10-11, 21, 31.

[16] Brown, *First Locomotives in America,* p. 211.

[17] *Ninth Annual Report, B&O* (1835), p. 27.

[18] *Tenth Annual Report, B&O* (1836), p. 9.

[19] *Eleventh Annual Report, B&O* (1837), p. 19, Appendix D.

[20] G. W. Whistler, Jr., *Report Upon the Use of Anthracite Coal in Locomotive Engines on the Reading Railroad* (Baltimore, 1849), p. 27.

and Ohio had purchased several wood-burning engines, which it used chiefly to haul freight; but the road still could boast of eleven anthracite locomotives.[21] The survey, therefore, proved one significant fact—that no railroad in the country burned raw bituminous coal as locomotive fuel.

The interest of the company drove it west and north following the winding ribbon of the Potomac to the Frostburg field in Allegany County, Maryland, the site of the Cumberland bituminous region. It was estimated that the fuel could be purchased for two dollars per ton or less should the line gain access to the area. The lure of low price was worth the risk of experimentation, for should Cumberland bituminous prove to be a fitting substitute for anthracite, it would furnish "perhaps the cheapest combustible which the earth produces."[22]

In 1837, when the Baltimore and Ohio began its first experiments with bituminous coal, it again assumed the position of a pioneer in fuel experimentation. It has been pointed out that not one road in the nation burned raw bituminous coal in its locomotives. The Philadelphia and Columbia Railroad used a mixture of Pittsburgh coal and wood and should not be considered an exception to this statement. The Pittsburgh bituminous, due to transportation charges, was expensive, and the railroad paid as high as thirty cents per bushel or about nine dollars per ton. Wood could be purchased for $4.25 per cord, stated the Baltimore and Ohio survey. On the seventy-seven-mile run between Philadelphia and Columbia, a locomotive burned an average of two cords of wood and fifteen bushels of Pittsburgh coal.[23]

The Baltimore and Ohio seemed undaunted by these figures. With wood from the Maryland area costing in the neighborhood of three dollars per cord, and with the prospect of cheap bituminous coal becoming available when the line penetrated the western Maryland fields, the company proceeded with its fuel experiments. There was tacit admission, too, concerning some of the difficulties experienced in using anthracite. Over the years burned-out grate bars and fireboxes had been particularly troublesome. These were problems which were to confront another great railroad which pierced the heart of the Pennsylvania anthracite regions. But the chief reason for the eventual

---

[21] Baltimore and Ohio Minute Book, September 11, 1837; November 7, 1838.
[22] *Eleventh Annual Report, B&O* (1837), p. 21, Appendix D.
[23] *Ibid.*, pp. 28-30.

abandonment of anthracite fuel by 1840-41 was the economy of bituminous coal once Cumberland would be reached. By 1841 the Baltimore and Ohio found that the trials over an extended period indicated the following ratio:[24]

1 ton of anthracite—1.25 tons of Cumberland bituminous
1 ton of Cumberland bituminous—1.87 cords of wood

The first fuel experiments burning Cumberland bituminous began on the old "Arabian," an anthracite locomotive with a vertical boiler. The inefficiency must have been apparent, but the trials were continued on the other ten anthracite locomotives, with no record of drastic mechanical alteration. When Cumberland was reached in 1842 it was a certainty that anthracite would be discontinued as fuel. Within a year the line had extended a spur into the heart of the mining region, a few miles above the town. The soft coal was then obtained for as little as one dollar per ton. At the same time anthracite at Baltimore was priced at $6.50 per ton. The company then bent all efforts toward the perfection of bituminous-burning locomotives.[25]

The transition was accomplished with difficulty. It was not until 1844 that Ross Winans built the first "successful" bituminous-burning locomotive with a horizontal boiler. Since anthracite had been rejected, the Baltimore and Ohio continued to wrestle with the problems of the new mineral fuel. Meanwhile, the company burned quantities of wood on the Washington extension and west to Harpers Ferry. In 1848 the Baltimore and Ohio's fifty-seven locomotives burned more wood than coal. The company complained bitterly of rising wood fuel costs.[26] The dream of exclusive use of cheap mineral fuel had been shattered by technical difficulties in its application to locomotive steam power. Smoke, clinker, sparks, and mechanical damage due to intense heat and soot plagued the chief engineer. The determination to surmount these obstacles was more than admirable. By the mid-fifties most of the major fuel problems had been solved. The railroad, connecting with Pittsburgh, moved through two hundred miles of bituminous coal country, much of which was within the bounds of Pennsylvania. The line continued to favor the

---

[24] Whistler, *Report*, p. 23.
[25] *Sixteenth Annual Report, B&O* (1842), pp. 17-19; *Eighteenth Annual Report, B&O* (1843), p. 12.
[26] *Twenty-second Annual Report, B&O* (1848), pp. 27, 36.

Cumberland coal of Maryland, however, and in 1856 consumed more than seventy-five thousand tons in its locomotives, paying an average rate of only eighty cents per ton delivered.[27]

A quarter of a century had elapsed between the time when the first practical anthracite engine was put into service on the Baltimore and Ohio and the extensive use of raw bituminous coal in its locomotives. During that time other railroads had been busy with fuel experiments. With the exception of the Philadelphia and Reading, the most significant road to engage in early experimentation with Pennsylvania coal was the Philadelphia and Columbia, a State-controlled line about eighty miles in length. In 1835 this railroad possessed seventeen locomotives, most of which had been built by the Baldwin Locomotive Works of Philadelphia, and which were specifically designed to burn wood. Soft coal or coke usually was mixed with wood and the result, other than the expense, seemed satisfactory. The fuel expenditure for 1835, taken from the Pennsylvania Canal Commissioner's Report was:[28]

$6,286.50—wood @ $4.55 per cord
    342.50—coke @ .25½ per bushel—1337 bu.
    322.50—bituminous coal @ .22½ per bushel—1430 bu.
    273.00—bituminous coal @ $5.25 per ton

More than six times the amount of money was spent on wood as on Pennsylvania coal. The coal interests in the State legislature were puzzled. Investigation resulted, followed by resolutions professing concern for the public welfare. Farmers along railroad lines had complained from time to time to the legislature that the shower of wood sparks from the stacks of the locomotives set fire to barn roofs and fields. The first Senate resolution began on this very note, but ended in a rather ludicrous attempt to legislate technological advancement:

> Whereas, the use of wood for fuel on the Railroads of this Commonwealth is productive of danger, and occasions much apprehension to the owners of property through which such railroads pass, which might be avoided by the use of mineral coal;
>
> Therefore,
>
> Resolved, that the committee on Roads, Bridges and Inland Navigation inquire into the practicability and expediency of

---

[27] *Twenty-ninth Annual Report, B&O* (1855), pp. 22-23; *Thirtieth Annual Report, B&O* (1856), p. 10.
[28] *Pennsylvania Senate Journal,* 1835-36, Appendix to II, 37.

using mineral coal, exclusively, as fuel for locomotives on the railroads of this Commonwealth, and of prohibiting, by law, the use of any other fuel for such purpose.[29]

The resolution was adopted after a second reading on December 18, 1837.[30]

Governor Joseph Ritner had taken a stand against State experiments with mineral coal for locomotive fuel. With the Panic of 1837 and an election year approaching, he cried that such experiments would be too expensive for the Commonwealth to bear. Instead, Pennsylvania should encourage individual enterprise to conduct the trials.[31] The resolution carried, nevertheless, and the Board of Canal Commissioners was directed to carry out fuel experiments on the Philadelphia and Columbia. In 1838 the report was forwarded to the Senate. Instead of coke or bituminous coal, anthracite was used in a newly purchased anthracite locomotive built by Ross Winans, of Baltimore.[32] The Motive Power Department was quite encouraged and stated that economy and saving would be effected with the substitution of Pennsylvania anthracite for wood. The report concluded with the following remarks:

> The experiments which the board directed to be made, have established the fact that anthracite coal can be successfully used as fuel in propelling, at any required speed, the locomotive engine, while its use will add to the security of the passengers, and the safety of the property of persons bordering upon the road.[33]

Ritner rode with the prevailing political breezes, and in his last annual message declared that the Philadelphia and Columbia Railroad had had "complete success" in using anthracite. He stressed saving and safety as his major themes.[34] At the very time that the Baltimore and Ohio began its experiments with Maryland bituminous coal and wood, and began gradually abandoning Pennsylvania anthracite, the State railroad in Pennsylvania rejected wood, became budget conscious over the high freight costs of Pennsylvania bituminous, and proclaimed the triumph of hard coal in locomotive engines. The victory was announced prematurely—by nearly two decades.

---

[29] *Ibid.*, 1837-38, I, 57.
[30] *Ibid.*, p. 103.
[31] *Pennsylvania Archives*, Fourth Series, VI, 392-393.
[32] *Pennsylvania Senate Journal*, 1840, Appendix to II, 71.
[33] *Ibid.*, 1838-39, Appendix to II, 7.
[34] *Pennsylvania Archives*, Fourth Series, VI, 457.

It soon became obvious that the State of Pennsylvania under the newly elected Porter administration was making every effort to popularize the use of mineral coal in locomotive engines. In the beginning the new administration adopted a new approach. Expense was regarded as secondary because it was felt that money expended on fuel or new equipment would be well spent should experimentation prove the superiority of Pennsylvania coal.[35] Success meant prestige for the "Keystone State" and the Porter Democrats, as well as profit for the coalmasters. The Superintendent of Motive Power of the Philadelphia and Columbia, James Cameron, made glowing reports on the "most perfect success" attained in the use of mineral coal as fuel for the railroad's new locomotives, an anthracite-burner built by Ross Winans and two bituminous coal-burners constructed by the Norris Locomotive Works in Kensington.[36] The first engine built by the Norris Works dated back to 1831. It was an anthracite-burner which was tested on July 4, 1832, on the New Castle and Frenchtown Railroad. It proved a failure as the grate and fire surfaces were too small to produce sufficient steam power. The locomotive would run a mile or so, exhaust its steam, and stop short until a new supply was generated. Future designs were worked out and one of the first efficient locomotives for ascending steep grades with loaded cars, the "George Washington," was built by these works in 1836. This excited the attention of England, and in 1837 the Gloucester and Birmingham Railway in Britain purchased seventeen locomotives from this company.[37] These engines had to be bituminous coal- or coke-burners as the British roads did not burn wood or anthracite. When the Philadelphia and Columbia wanted to acquire two new bituminous-burning locomotives, it therefore turned to the Norris Works rather than to Baldwin's company which, at this time, specialized in wood-burners and had sold a score of them to the State railroad. Baldwin was proud of the wood-burners and even argued the virtues of forest fuel over raw coal before a Pennsylvania House committee.[38] The State railroad also purchased locomotives from D. H. Dotterer and Company of Reading, a firm which claimed its engines could burn either anthracite or wood.[39] This claim was exaggerated, evidently, as the firm went out of business in 1842.

---

[35] *Pennsylvania Senate Journal*, 1840, Appendix to II, 74; 1841, p. 76.

[36] *The Pennsylvanian*, February 25, 1840.

[37] Freedley, *Philadelphia and its Manufactures*, pp. 309-310.

[38] *Pennsylvania House Journal*, 1836-37, II, 821.

[39] *The Pennsylvanian*, September 17, 20, 1840.

In order to convert the wood-burning Baldwin locomotives to coal-burners, a plan was tried which had originated with the firm of East-wick and Harrison. It was a disappointment to the Philadelphia and Columbia. Eastwick and Harrison was the same Philadelphia concern which was induced by Czar Nicholas I to move to Russia in 1844 and build a large locomotive works on the fringes of St. Petersburg. By 1845 the new Russian works employed over thirty-five hundred men, with the skilled mechanics drawn from the British Isles, the Germanies, and the United States.[40] When Cameron tested the East-wick and Harrison plan, he found that it was designed to place the steam chest over the exhaust pipes. The steam drove the air out of the exhaust and created a vacuum into which outside air rushed to provide a draft. This idea worked well enough on level track or on the first ascent, but when the steam power was shut off on the descending grade, the locomotive arrived at the bottom of the slope with a low fire in the firebox insufficient to drive the train up the next grade. The locomotive came to a halt, then spent many minutes getting up enough steam to make the climb. Cameron solved the dilemma by attaching a small rotary steam blower to provide an artificial draft. This was kicked in as the engine reached the bottom of the hill, furnishing the necessary blast of air which heated the coals and produced a head of steam for the ascent. He later claimed that Winans' blower was superior to his and urged the State to buy the patent and apply it to the other engines on the road. With his own invention, however, Cameron claimed he could burn Schuylkill anthracite in the Baldwin locomotives. In 1839-40 Cameron reported five locomotives burning Pennsylvania anthracite and four burning Pennsylvania bituminous from the West Branch of the Susquehanna River. The following year all thirty-eight had been altered to burn anthracite or bituminous coal "exclusively." Cameron asserted that the practicability of Pennsylvania coal had been established and the objective of the State accomplished. Whether coal would be cheaper than wood, or as cheap, was still unknown, however, because of the large amounts of fuel wasted by the firemen in the improper firing of the boilers.[41] To this, one would have to add that the heat of the coal, anthracite or bituminous, burned out grate bars, clogged tubes, and destroyed fireboxes much more quickly than wood. This, of course, necessitated expen-

---

[40] *London Mining Journal.* Reprinted in *Fisher's National Magazine,* I, September, 1845, pp. 380-381.

[41] *Pennsylvania Senate Journal,* 1840, Appendix to II, 74; 1841, p. 76; *The Pennsylvanian,* February 25, 1840.

sive repairs which added to the overall cost and maintenance of coal-burning locomotives.

Cameron was entirely too optimistic, as the report of his successor showed in 1843. The Winans' vertical boiler-type locomotive was considered satisfactory, but not completely perfect, for anthracite coal. Most of the locomotives on the Philadelphia and Columbia were wood-burners with horizontal boilers which could not be converted to function efficiently with anthracite. The wood-burners could be used to burn bituminous coal when a blower was attached, but this arrangement also needed further experimentation. The Commonwealth now decided not to spend additional funds to purchase Winans' locomotive, and to economize on fuel. The railroad settled back to its old pattern of burning a mixture of wood and raw bituminous coal. Wood became scarcer and more expensive over the years, but the supplies of Pennsylvania soft coal increased and were cheapened through better transportation facilities. Although the Superintendent of Motive Power predicted that the locomotives would be able to burn raw bituminous exclusively without expensive alteration of parts,[42] the line by 1845 was spending over eighteen thousand dollars for soft coal and nearly twenty-five thousand dollars for wood to burn in its locomotives.[43]

The decision to desert anthracite did not solve the problem of flying sparks. In spite of spark arresters, bits of burning wood and pieces of glowing bituminous embers poured out of the smoke stack each time the locomotive was fired with fuel. The bituminous embers endured, held their heat, and were more dangerous than wood sparks.[44] The fire hazard became a serious consideration for the legislature, which finally passed a law on February 3, 1846, which provided compensation to those who suffered fire loss caused by flying sparks from the State's locomotives. The Commonwealth even paid to move barns a safe distance from the right-of-way.[45]

In the late forties and the succeeding decade, experimentation with Pennsylvania coal as locomotive fuel was resumed on the Philadelphia and Columbia and on the other lines within and outside the State. One railroad was outstanding in its contribution to locomotive fuel experimentation. This was the Philadelphia and Reading, which reached the anthracite regions in 1842 and within five years became

---

[42] *American Railroad Journal,* XVI, May, 1843, pp. 133-134, September, 1843, p. 281.

[43] *Pennsylvania Executive Documents,* 1845, p. 11.

[44] *American Railroad Journal,* XVI, May, 1843, p. 135.

[45] *Pennsylvania Executive Documents,* 1846, pp. 86-87.

the nation's leading freight line. As Reading locomotives hauled their endless trails of coal cars to tidewater over the ninety-odd miles of track, it seems unbelievable that for more than a decade these iron-wheeled steam machines burned wood in their fireboxes.

THE READING LINE

The opening of the Philadelphia and Reading Railroad Company line north through the valley of the Schuylkill to Mount Carbon, just below Pottsville, on a cold January day in 1842 was the occasion for great celebration. Guns were fired in salute and cheers were raised as the train, forty-three cars long, left the depot carrying twelve hundred citizens from Pottsville and vicinity, three military bands, and detachments of the militia. Below Mohrsville another locomotive followed, drawing fifty-two coal cars. Several hundred persons joined the others at Reading. When the trains arrived at Philadelphia, the more than two thousand passengers paraded through the streets with bands playing. Banners, bearing inscriptions, were outlined against the wintry sky. One slogan proclaimed the objective of the company's extension into Schuylkill County: "We penetrate the mountains to bring out treasures to add to your comfort and prosperity."[46] It was the attraction of the coal trade of the anthracite fields which had drawn the railroad north from Reading to Pottsville. Two years before, the southern section had been completed to Philadelphia. Now an iron trail of ninety-four miles of single track stretched from the rough hills of Schuylkill County to Port Richmond on the Delaware. The purpose of the company was clear: the profitable freight was to be anthracite coal. By a peculiar twist of technological irony, anthracite coal was to move to tidewater drawn for several years by Reading locomotives burning wood fuel.

The use of wood was not considered even a remote possibility in the beginning. The Baltimore and Ohio had burned anthracite for years with apparent success, and the Philadelphia and Columbia seemed about to follow that example. With the establishment of a coal railroad which bored into the core of the richest anthracite area in the world, Schuylkill County, it would be absurd to think of any other fuel for steam power.

Experiments with anthracite began in 1838 when the Reading extended only as far as Pottstown. The first steam locomotive to run on the new track was an anthracite engine designed by Moncure Robinson,

---

[46] Hazard, *U. S. Register*, VI, January 26, 1842, p. 63.

famed builder of the Reading, and constructed in the shops of East-wick and Harrison, of Philadelphia. The locomotive was appropri-ately named the "Gowan and Marx" after the British banking house which had financed the building of the road.[47] When the line was opened to Philadelphia in December, 1839, the "Gowan and Marx" and the "Delaware," a Winans anthracite locomotive, were exhibited by the company. Moncure Robinson seemed certain that the fuel of the Reading locomotives would be anthracite coal.[48] He had high praise for the "Gowan and Marx," as should be expected since he was the designer. In 1839 Robinson reported to the company that "her draft and generation of steam with anthracite coal, appears to us as perfect as in any locomotive that we know of."[49] The Reading adver-tised that among many advantages the railroad would have over the Schuylkill Canal would be reasonable freight charges made possible by the use of cheap anthracite coal as locomotive fuel.[50]

Technical problems of design and metallurgy added to mainte-nance costs and disrupted the plan to use anthracite coal as locomotive fuel. Control and retention of steam produced by the slower-burning anthracite, the translation of steam power to adhe-sion of wheel to track for heavy loads, long hauls and steep ascents, as well as incomplete combustion all presented themselves for solu-tion. The heat of the hard coal also blistered the iron sheets of the firebox. Copper was substituted, but the sharp bits of coal scored the softer metal. Sometimes the grate bars melted through improper firing as the accumulation of cinder and ash cut off the cooling effect of circulating air. Repair costs were heavy and time was lost in the shops. The estimated yearly maintenance cost for a coal-burning locomotive was four to five times that of a wood-burner.[51] These were major problems which manifested themselves almost as soon as the trial runs began. They were to remain major problems for well over a decade. The peculiarity of the situation was evident. Anthra-cite was cheap fuel to the Reading, but the maintenance of equipment

[47] W. W. Rhoads, *When the Railroad Came to Reading! A Newcomen Address* (New York, 1948), pp. 16-17.

[48] *Report of the Engineers of the Philadelphia and Reading Railroad Company with Accompanying Documents, etc.* (Philadelphia, 1838), p. 4.

[49] *Ibid.*, 1839, p. 7.

[50] "X," *The Reading Railroad: Its Advantages for the Cheap Transportation of Coal, As Compared with Schuylkill Navigation and Lehigh Canal* (Philadelphia, 1839), p. 17.

[51] Whistler, *Report*, pp. 18-19; *Journal of the Franklin Institute,* LI (January-June, 1851), 138.

necessitated by the destructive effects of anthracite forced the coal railroad to revert to wood during the early years. This was true, despite the availability of anthracite and bituminous, of much of the railroad operation throughout the United States prior to the Civil War. Yet scarcity and growing costs of wood and labor forced further experimentation even after the early failures. The Baltimore and Ohio, and later the Pennsylvania Railroad, led the way in the use of raw bituminous coal, whereas the Philadelphia and Reading was the major concern which experimented with anthracite fuel.

Between 1838 and 1843 the Reading gradually accepted the fact that the few coal-burning locomotives on their road were failures and alterations had done little or nothing to improve them. All but one of the company's thirty-nine locomotives burned wood by 1844. The report for that year sounded disheartened over anthracite experiments: "Several previous attempts to burn this fuel with advantage have been attended with an expense and inconvenience which in some cases, deranged the business of the road." A single-track railroad could ill afford frequent breakdowns in experimental coal-burners.[52]

Disheartened but not defeated the company turned to its own shops at Reading. These had belonged to the locomotive manufacturing firm of Dotterer and Company, which sold out to the Reading in 1842. While locomotive design and construction were studied at Reading, the locomotives in service on the road continued to burn wood.[53] Between 1844 and 1846 the number of locomotives had grown from thirty-nine to seventy-two. Most of the large new ones were Baldwin wood-l urners.[54] By 1847, the company, despairing over rising wood costs and heartened by encouraging reports on the new Winans anthracite-burners, purchased four of these Baltimore locomotives. At the same time the Reading shops under G. A. Nicolls put into operation what was thought to be the answer to most of the anthracite problems in coal-burning locomotives.[55] This was the "Novelty," a large locomotive built in two sections, one section housing the machinery and the other a large boiler and firebox. The locomotive was poorly designed, lacking adhesion and proper steam power. After repeated failures it was withdrawn from the road and,

---

[52] *Report of the President and Managers to the Stockholders of the Philadelphia and Reading Railroad Company* (1844), p. 12. Referred to hereafter as *Annual Report, Philadelphia and Reading*.

[53] *Ibid.*, 1845, p. 13; 1846, p. 23.

[54] *Ibid.*, 1847, p. 30.

[55] *Ibid.*, 1848, p. 15.

ironically enough, used to operate a sawmill. The new Winans locomotives were much better investments. Because the cab was built over the boiler, these locomotives presented a peculiar humped appearance and were promptly dubbed "Camels" by railroad men. The main complaint about the "Camel" locomotives was that their grates burned out too rapidly. Once again repair costs and lost time offset the saving in anthracite fuel. Prior to 1850 the company spent much more money on anthracite for use in stationary steam engines in its shops than for use in locomotives.[56]

In 1848 James Millholland became master of machinery for the Reading Company. A man in his middle thirties, he had worked since youth with locomotive and marine engines.[57] His experience and inventive genius served the railroad well. Millholland, more than any other individual, was responsible for solving the riddle of the proper and efficient use of anthracite coal in Reading locomotives. In 1851 and 1852 he patented boiler designs which were one-third more efficient than Winans'.[58] For the first time in ten years, the reports of the company sounded a hopeful note concerning the fuel problem.[59] During the fifties the costs of wages and maintenance rose, but the cost of transporting coal, merchandise, and passengers dropped. The reason was simple: sixty per cent of the transportation by 1854 was by thirty-one locomotives using anthracite coal.[60]

By 1856, eighty-five per cent of the coal brought to tidewater by the Reading was hauled by anthracite-burning locomotives.[61] The conversion of wood-burners to coal-burners continued in the Reading shops. The company also added to the number by building new locomotives in its shops and by purchasing Winans locomotives and making its own improvements. One improvement was a firebox with inclined or sloped copper sides. The angle prevented the scoring of the metal by the coal when the boiler was charged with fuel. Millholland was responsible for this innovation. A most significant improvement, again introduced by James Millholland, eliminated costly repairs of burned-out grate bars. The invention of the wrought-iron tubular water-cooled grate not only reduced maintenance costs to a minimum, but saved fuel. In order to give longer life to the old cast-

---

[56] *Ibid.,* 1849, p. 23; *A Century of Reading Company Motive Power* (Philadelphia, 1941), p. 22; see Table E, Appendix.

[57] *A Century of Reading Company Motive Power,* p. 24.

[58] *Miners' Journal,* October 4, 1851.

[59] *Annual Report, Philadelphia and Reading* (1853), p. 26.

[60] *Ibid.,* 1854, p. 31.

[61] *Ibid.,* 1855, p. 28.

iron grate bars, the Reading had been burning a gray-ash anthracite, softer and a little cheaper than the superior Schuylkill white-ash. Gray-ash anthracite did not give off as much heat as white-ash and, consequently, grate bars lasted a little longer. Unfortunately, the small advantage in price and longer lasting grates were offset by the impurities contained in the gray-ash coal. Incomplete combustion, clinker, and ash added to fuel costs. With Millholland's water-cooled grates it was possible to burn the white-ash anthracite and save money on fuel.[62]

The Reading continued to burn wood in some of its passenger engines to 1860, but by 1865 conversion to coal was almost complete. Some wood still was used to ignite the coal when the locomotive boilers were "fired up," but the ring of the axe, the sound of the steam saw, and the sight of great piles of timber, cut, stacked, and ready for the wood agent's appraisal, had all but ceased along the line of the Reading by the end of the Civil War. The key to the utilization of anthracite coal in locomotive engines belongs to the Reading. No other railroad had more to gain by its application. No other line tried harder to solve the problem, and no other line so well deserved the success which came through persistence and ingenuity.

THE ROOTS OF TRANSITION

The last half of the eighteen-forties saw the formation of a railroad which was to grow into the largest system of rail transportation in the world. The Pennsylvania Railroad issued its first report to the stockholders in 1848. Four years later it had tunneled through the Alleghenies and linked Pittsburgh with Philadelphia. Unlike the Reading, it paid a heavy tax to the State for it was to compete with the State Works until the latter was entirely sold or discarded. Before the Civil War the Pennsylvania had shaken the tax by purchasing much of the State Works, including the Allegheny Portage Railroad. The ponderous stationary steam engines and the inclined planes of one of the most interesting transportation systems in America were forgotten as the Pennsylvania Railroad Company drove through the mountains to the West.[63]

Moving across the State, the Pennsylvania Railroad invaded the rich bituminous coal fields. Between 1849 and 1851 the road burned

---

[62] *Ibid.*, 1859, p. 22; 1860, pp. 55-58.

[63] R. S. Elliott, *Notes Taken in Sixty Years* (St. Louis, 1883), p. 83.

wood, and in that short period spent over twenty-five thousand dollars for wood fuel and only $547 for coal.[64] But gaining the terminus at Pittsburgh, the line began a concentrated effort to burn mineral fuel in its locomotive engines. Experiments were conducted with coke and Pittsburgh and Allegheny coal. The latter was a general term commonly applied to bituminous coals, other than Pittsburgh, which came from the Great Allegheny Field. The chronic difficulties of burned-out grates and fireboxes plagued the Pennsylvania, as bituminous coal, though not as destructive as anthracite, threw off more heat than wood and damaged the locomotive. A mixture of wood and soft coal alleviated the repair problem, but did not solve fuel costs.[65] Coal experiments were continued and in 1854 the company, which kept very incomplete records of its fuel tests, purchased thirty-five new engines, twenty-one of which were coal-burners.[66] Two years later the costs of wood and coal on the "Main Line" were $49,000 and $52,000 respectively.[67] This indicated progress, but also emphasized that wood still was a major part of the railroad's fuel, and much remained to be accomplished before locomotives on the Pennsylvania burned raw bituminous under their boilers. The chief fuel problem of the railroad in the late fifties seemed to have been the smoke from bituminous coal. This was true particularly of Pittsburgh coal. It was not until 1860 that the Master of Machinery, George Grier, reported that the smoke nuisance had been solved. Praise was given justly to the firm of Gill and Company for the perfection of a smoke-consuming apparatus, but much of the credit should go to an energetic young man who spent tedious weeks on the western division experimenting with equipment and coal.

This young man was William Jackson Palmer, a native of Delaware, who was to rise to the rank of general in the Union Army during the Civil War, then move west after Appomattox and become president of the Denver and Rio Grande Railroad, the Colorado Coal and Iron Company, and finally the Rio Grande and Western Railroad. In the summer of 1855, when only nineteen, Palmer traveled to England, paying part of his expenses by writing articles on British

---

[64] *Fourth Annual Report of the Directors of the Pennsylvania Railroad Company to the Stockholders* (1851), p. 58. Referred to hereafter as *Annual Report, Pennsylvania Railroad.*

[65] *Fifth Annual Report, Pennsylvania Railroad* (1852), pp. 41-42.

[66] *Seventh Annual Report, Pennsylvania Railroad* (1854), p. 53.

[67] *Tenth Annual Report, Pennsylvania Railroad* (1857), p. 55.

coal mining for Benjamin Bannan of the *Miners' Journal.* Gerard Ralston brought these letters to the attention of J. Edgar Thomson, president of the Pennsylvania Railroad, who urged Palmer to find out as much as he could about English methods of burning coal and coke in locomotives.[68] The young Palmer threw himself into the investigation with his customary enthusiasm. But he did not supply Thomson with any new or practical information, other than his observation that the principal problem was smoke consumption and that the British had all but solved it by shifting to coke or by patenting smoke-consumers.[69] While in England, Palmer became interested in the calcination of Welsh anthracite. This was really a special coking process in a closed oven or retort which drove off the small amount of volatile material contained in anthracite, leaving almost pure carbon. According to the inventor, Charles Morgan, calcinated anthracite would make the best locomotive fuel in the world.[70] Palmer wanted Morgan to make him his agent in the United States. The young man told Morgan that he had connections with the leading anthracite journal of the nation as well as with the *New York Tribune.*[71] Morgan seemed pleased with the opportunity of introducing his process to America, but when Palmer returned home with the idea in the fall of 1856, he got a cool reception. Morgan persisted in writing letters, but finally became discouraged when he learned of the $400 patent fee. He then asked Palmer to introduce the process as his own idea and enclosed directions.[72] Palmer evidently considered the whole plan unacceptable to American railroads for he busied himself with other things. He became secretary to the Westmoreland Coal Company and helped to introduce the excellent western Pennsylvania gas coal to eastern cities. On June 1, 1857, Palmer, whose first love was the railroad, became confidential secretary to J. Edgar Thomson, president of the Pennsylvania.

It was at this time that the Pennsylvania Railroad intensified its campaign to burn raw bituminous and coke in its locomotives. The Westmoreland Coal Company was supplying coal to the road at eighty-

---

[68] J. S. Fisher, *A Builder of the West, the Life of General William Jackson Palmer* (Caldwell, Idaho, 1939), p. 34.

[69] CSHS, Palmer Papers, W. J. Palmer to J. E. Thomson, September, 1855, June 19, 1856.

[70] *Ibid.,* Charles Morgan to W. J. Palmer, February 9, 1856; Charles Morgan to Gerard Ralston, November 8, 1856.

[71] *Ibid.,* W. J. Palmer to Charles Morgan, n.d. [ed. 1856].

[72] *Ibid.,* Charles Morgan to W. J. Palmer, February 20, 1856, April 27, 1857.

five cents per ton and coke at four cents a bushel.[73] The result was not too satisfactory. By the spring of 1859 Palmer found himself entrusted with the responsible job of testing the new Phleger coal-burning boiler, a new Winans locomotive, and a dozen inventions in grates, fireboxes, smoke-consumers, and plans for firing boilers.[74] The two types of coal which proved most economical and efficient for the long run across Pennsylvania were Pittsburgh bituminous and Broad Top semibituminous.[75] Broad Top coal was found in a partially isolated sector of the State east of the Great Allegheny Field and between Bedford and Lewistown. The area was connected with the Pennsylvania Railroad by the Broad Top and Huntingdon line. Palmer, soot-stained and determined, rode locomotive engines for weeks, making careful observations and taking detailed notes for his report. The people in central and western Pennsylvania were so intensely interested in the railroad that Palmer, writing to his friend Isaac Clothier, of Philadelphia, at a time when the nation was in the midst of sectional tension, said, "If Beecher should go to Altoona, he would find himself without a subject, unless he chose 'Motive Power.' "[76]

It was Palmer's experiments and observations which led the Pennsylvania Railroad to report in 1860 that Pennsylvania bituminous coal could be burned in all passenger engines and freight engines "with great economy" and without the usual smoke. The Master of Machinery then recommended the alteration of all wood-burning locomotives to coal-burners "as fast as circumstances will permit."[77]

While the Pennsylvania experienced success with raw bituminous and semibituminous coal and moved rapidly toward conversion to mineral fuel, the Baltimore and Ohio, burning the raw Cumberland bituminous and wood, began experiments with coke in 1854. These experiments were directed toward its passenger engines, which, since the abandonment of anthracite, had been burning wood. Soft coal presented a smoke and cinder hazard to passenger trains which was not of much consequence on freight lines. Wood, though expensive, did not throw off as much smoke as bituminous coal. Coke proved

---

[73] *Ibid.*, Coal Notes, August 1, 1856.

[74] *Ibid.*, Palmer Notes, Coal Burning Engine, 1859.

[75] *Ibid.*, "Report on Pittsburgh and Broad Top Coals of the Pennsylvania Railroad, February 1, 1860."

[76] I. H. Clothier (comp.), *Letters 1853-1868, General William J. Palmer* (Philadelphia, 1906), p. 15.

[77] *Thirteenth Annual Report, Pennsylvania Railroad* (1860), p. 27.

to be cheaper than wood, but not so cheap as raw coal. The ratio was recorded as follows:

wood—7.8¢ per mi.
coke —5.6   "   "
coal —3.6   "   " [78]

With new smoke-consumers, Baltimore and Ohio passenger engines burned either coke or raw coal. By the time of the Civil War the Baltimore and Ohio was pursuing the last phase in becoming a coal-burning railroad. The process had taken a generation.

The three great railroads, the Reading, the Pennsylvania, and the Baltimore and Ohio, expended large sums of money in fuel experimentation. These lines, with the State-controlled Philadelphia and Columbia, which became a part of the Pennsylvania Railroad system, led American railroads in the transition from wood to coal. Many other lines conducted experiments but, lacking capital, did not have the success of these three. New England railroads were pressed sorely for economy in fuel consumption, but most of them delayed experimentation until the fifties when newer inventions made the tests better financial risks.[79] The Philadelphia, Wilmington and Baltimore burned only thirty-two hundred tons of coal to over eighty-five hundred cords of wood as late as 1859.[80] In western Pennsylvania, in the heart of the bituminous coal fields, the Cincinnati and Pittsburgh Railroad had only two coal-burners in operation in 1855. In 1858 twenty locomotives used wood and twenty-two burned half wood, half soft coal, and the ratio remained the same two years later.[81] The Pittsburgh, Fort Wayne and Chicago Railroad was spending more than eighteen thousand dollars for coal fuel and nearly sixty-eight thousand dollars for wood on the eve of the Civil War.[82] In the anthracite sections of Pennsylvania the Beaver Meadow and Hazleton roads had succeeded quite early in burning anthracite on their short runs. The Lehigh Valley also could be classified as a successful coal-burning

---

[78] Mendes Cohen, *Report on Coke and Coal Used for Passenger Trains on the Baltimore and Ohio Railroad* (1854); *Thirty-second Annual Report, B&O* (1858), p. 12.

[79] *Miners' Journal*, June 23, 1855; see CSHS, Palmer Papers, pamphlets, *Boardman's Coal-Burning Locomotive Boiler Company* (1856) and F. E. Felton, *Mineral Fuel for Locomotives* (Philadelphia, 1857).

[80] *Pennsylvania Legislative Documents* (1860), p. 572.

[81] *Annual Report, Cincinnati and Pittsburgh Railroad* (1856-1860).

[82] *Annual Report of the Pittsburgh, Fort Wayne and Chicago Railroad* (1858-1859).

line by the end of the period, but most of the minor coal-carrying railroads in the eastern part of Pennsylvania continued to use wood fuel to fire the boilers of their older equipment.[83] From the tiny anthracite-hauling railroad, the Swatara and Tioga, which burned more wood than coal, to the 288-mile Sunbury and Erie and the New York and Erie, both of which burned wood and only wood in their locomotives as late as 1860, a similar pattern could be repeated throughout the country.[84] With few exceptions, wood fuel still was dominant in the nation's nearly nine thousand locomotives.

For all that, the few decades preceding the Civil War were definitely years of transition and growth, when economic pressure, technical change, and inventive genius led to the development of the coal-burning locomotive in America.

---

[83] See Table F, Appendix; *Miners' Journal,* September 4, 1847.
[84] *Pennsylvania Legislative Documents* (1861), pp. 592-594, 601; Table E, Appendix.

CHAPTER VII

The Expansion of Markets

THE ANTHRACITE TRADE

B ETWEEN 1820 and 1860 references to "the coal trade" in newspapers, journals, and other publications usually pertained to the commerce in Pennsylvania anthracite. Over a span of forty years the Pennsylvania anthracite coal trade increased from 365 tons brought to Philadelphia by way of the Lehigh Navigation in 1820 to a reported more than eight million tons mined and shipped to markets in 1860. By the Civil War more than four million tons were mined annually in Schuylkill County alone. This amount was one million tons in excess of the bituminous production of the State and half a million tons greater than the total bituminous produced in the other fifteen bituminous-producing states and territories. The coal production of the United States in 1860, estimated by the *Eighth Census,* may be noted at a glance.[1]

	Anthracite-Tons	Bituminous-Tons	Totals
Pennsylvania	8,114,842	2,934,512	11,049,354
All others—U.S.	1,000 (Rhode Island)	3,283,568	3,284,568
Totals	8,115,842	6,218,080	14,333,922

The principal explanation for the increase in anthracite production is found in the demand for the varied uses of the mineral fuel in home, factory, blast furnace, and steamboat. In order to supply the growing markets large amounts of capital were invested; numerous mines, controlled by individuals and by corporations, were opened; and transportation routes were developed to bring the fuel to the places of consumption.

The development of transportation routes from the Pennsylvania anthracite regions to tidewater, western New York, and the Great Lakes, as well as the extensive growth of coastwise commerce out of Philadelphia, was devoted primarily to bringing anthracite coal to market to meet increased demand. The evolution of the network of

---

[1] *Eighth Census, Manufactures,* pp. clxiii-clxiv.

water and rail routes coincided with the acceptance of anthracite fuel on an ever-increasing scale and the consequent growth and expansion of markets. The difficulties, financial and physical, faced by individuals, corporations, and the states of Pennsylvania, New York, and New Jersey in constructing and maintaining the arteries of the coal trade form detailed and separate chapters in the history of American transportation and commerce. But the general picture of coal transport was not set in a separate frame. It became a part of the whole scene, portraying the rise of an industry which, by the end of the Civil War, became basic to the comfort, economy, and manufacturing enterprise of the American nation.

Beginning in 1820 with the first successful commercial shipments of anthracite by way of the Lehigh Navigation, Philadelphia became a focal point of the coal trade. With the succeeding decades increased tonnages of anthracite entered the city over three well-defined routes: the Lehigh Navigation and the Delaware State Canal, the Schuylkill Navigation, and the Reading Railroad.[2] A portion of the fuel was consumed by Philadelphia's homes and industries. Even more was transshipped to New York and to other markets along the eastern seaboard. Philadelphia soon became the foremost coal port in the United States. From early spring to late fall a steady train of colliers hoisted sail, moved down the Delaware, and rounded Cape May for ports from Maine to the Carolinas. Through the anthracite coal trade Philadelphia tripled her coastwise tonnage. By 1860 more than four million tons, or approximately one-half the anthracite production of Pennsylvania, passed through the city on its way to markets in the United States and abroad.[3]

Early navigational improvements on the Lehigh from Mauch Chunk to Easton, where the river joins the Delaware, enabled crude pine arks laden with coal to pass from the anthracite regions of the Lehigh Valley to Philadelphia. For over a decade the transportation facilities furnished only descending navigation for the arks. After the great, cumbersome, box-like affairs reached tidewater and unloaded their cargoes, they were broken up and their pine planks sold for lumber. This was an inefficient and expensive procedure, forcing the Lehigh Coal and Navigation Company to employ large numbers of men to cut timber and build new arks each time a shipment of coal was sent

---

[2] *Pennsylvania Coal and Its Carriers* (Philadelphia, 1852). Essays reprinted as editorials from the Philadelphia *North American*.

[3] *Merchants' Magazine*, XVI, 202; XLVI (January-June, 1860), 583.

downstream. In 1829 ascending navigation on the Lehigh was completed from Easton to Mauch Chunk, but the State did not permit the company to improve the Delaware. Instead, Pennsylvania began its own project and supplied the necessary ascending navigation from Bristol to Easton by the Delaware Division of the Pennsylvania State Canal.[4] Work progressed slowly and the route was not completed until 1833. The State constructed the canal with the plan that the major source of income would be gained from the coal trade. Yet the commerce of the upper Delaware Valley was not overlooked. The Delaware Division would furnish an outlet for the products of the valley and open an era of prosperity to the inhabitants of the region.[5] The improvements on the Delaware permitted the Lehigh company to introduce permanent canal boats the length of the waterway from the mines to Philadelphia.

During the first five years of the trade, from 1820 to 1825, the Lehigh firm supplied the Philadelphia market on a modest scale. In possession of both mining and transportation privileges, the company did not have any competitor and sold directly to the consumer. After 1825 Lehigh coal competed with anthracite from the Schuylkill Valley. As anthracite became more popular and demand as well as competition increased, commissions were paid to persons who obtained fuel orders from manufacturing firms. It was sixteen years, however, before the company entirely relinquished its retail business in Philadelphia and dealt only in wholesale rates. In 1836, the same year in which the company abandoned its retail policy in Philadelphia, the first shipments of coal mined by independent concerns in the Lehigh region began moving down the canal to tidewater markets. By 1840 the amount of coal from these sources surpassed that mined by the Lehigh Coal and Navigation Company. Still, much of the profit accrued to the Lehigh firm, since it controlled the transportation route from the coal fields to Easton.[6]

From Easton, the coal which was to pass to the Philadelphia market was shipped over the Delaware Division of the State Canal. Despite its recognized inadequacies of narrow locks and low water, the Delaware Division, by the end of the period, was furnishing Philadelphia with over half a million tons of anthracite per year.[7] Not all this coal

---

[4] Shegda, "Lehigh Coal and Navigation Company," p. 103; C. L. Jones, *The Economic History of the Anthracite-Tidewater Canals* (Philadelphia, 1908), p. 17.

[5] *Pennsylvania Archives,* Fourth Series, V, 915-916.

[6] Shegda, "Lehigh Coal and Navigation Company," pp. 160-163, 179.

[7] *Twenty-sixth Annual Report of the Philadelphia Board of Trade* (1859), p. 146.

came from the Lehigh Valley. Some was drawn from the western end of the Wyoming field and was brought to market over the twenty-mile Lehigh Railroad from Wilkes-Barre to White Haven and then by slackwater navigation another twenty-six miles to Easton.[8]

When the Schuylkill County mines in the vicinity of Pottsville began producing coal for the eastern market, Philadelphia functioned as a center of consumption and, above all, of distribution to other points. Until 1842 the only major artery of commerce from the Pottsville area was by way of Schuylkill Navigation. The slackwater canal project was not undertaken with the coal trade as an object of special importance. Unlike the Lehigh company, the Schuylkill firm did not possess mining privileges. The canal was to become, nevertheless, the most significant coal route from the anthracite regions to tidewater until its tonnage was surpassed by the Philadelphia and Reading Railroad in 1844.[9] From 1825, when the first regular shipments via the Schuylkill Canal arrived in Philadelphia, until 1842 when the Reading reached Mount Carbon, the waterway enjoyed a transportation monopoly. By 1840 its coal shipments for Philadelphia's consumption totaled one hundred thousand tons. An additional quarter of a million tons descended to tidewater by the same route and were transshipped from Philadelphia wharves to New York and other urban centers along the coast.[10] Seven coal railroads intersected the canal at various points. One of these roads was the Little Schuylkill from Tamaqua to Port Clinton, a twenty-five-mile line founded by Friedrich List, German economist and political refugee, and financed by Stephen Girard, Philadelphia's famous merchant prince.

A narrow ribbon of water, the Union Canal, connected the Susquehanna and Schuylkill rivers. The canal never became a great coal carrier because of its shallow water and narrow locks. Strong arguments were presented to the State for funds to enlarge the waterway so that eastern markets could be supplied with Pennsylvania mineral resources. Both eastern anthracite and western bituminous coals would become profitable freight if the canal were improved.[11] A committee reporting to the legislature on the Swatara mining district

---

[8] Hazard, *U.S. Register,* I, September 4, 1839, p. 201.

[9] J. M. Sanderson, *A Letter on the Present Condition and Future Prospects of the Reading Railroad in January 1855* (Philadelphia, 1855) , p. 10.

[10] *Report to the Stockholders of the Schuylkill Navigation Company, January 6, 1840.* Reprinted in Hazard, *U.S. Register,* II, January 22, 1840.

[11] *Memorial of a Convention of Citizens of the Commonwealth for Aid to Enlarge the Union Canal* (Harrisburg, 1839) , pp. 3-13.

also recommended enlargement, claiming that anthracite passing over the Union Canal could not compete with Schuylkill Valley coal. The cost of transportation on the small twenty-five-ton boats of the canal exceeded the value of the coal itself.[12] A bill for State funds was passed by the legislature, but failed to become law after Governor David Rittenhouse Porter's veto. The Governor based his action on economy. It was not until 1856 that the canal was finally enlarged to pass boats of seventy-five to eighty tons. By that time railroads had begun to triumph over the "canal age," and the Union Canal, like other canals in the Commonwealth, eventually became obsolete.[13]

During the period from 1833 to 1847 the Union Canal, in spite of its shortcomings, functioned as the only outlet to eastern markets from the Pine Grove or Swatara district of the southern anthracite field. A "feeder" from the region cut into the main line of the Union Canal almost halfway between Middletown and Reading. The red-ash anthracite, prized by many for use in open grates in the home, was then floated east on the Union Canal and south on the Schuylkill to Philadelphia and the seaboard. In 1847 a railroad was built from the area to connect with the Reading line and furnish a second outlet to market. By 1850 anthracite shipments from Pine Grove via canal totaled only seventy thousand tons annually, but an amount estimated to be equal to that was brought to market by rail.[14]

The Union Canal and the railroad connecting the Swatara mines with the Schuylkill Canal and the Reading should be considered parts of the coal transportation system which supplied the port of Philadelphia and its domestic market. Narrow locks, shallow water, and small boats hampered the full development of the Union Canal's coal trade. Similar problems had confronted the Schuylkill Navigation Company as anthracite increased in tonnage and importance and contributed to most of the firm's profits. The company, straining every financial resource, deepened the canal and enlarged the locks. Boats of thirty to thirty-five tons were in use to 1835. Five years later boats averaged fifty tons, and by the late forties some had a capacity of 180 tons, the largest on any canal in the East. Most of the boats on the Schuylkill Canal, some costing as much as twenty-four hundred dollars, were owned by individual shippers or by mine operators who

---

[12] *Report to the Legislature of Pennsylvania Containing a Description of the Swatara Mining District* (Harrisburg, 1839), pp. 51-53.

[13] J. L. Livingood, *The Philadelphia-Baltimore Trade Rivalry, 1780-1860* (Harrisburg, 1947), pp. 110-114.

[14] Taylor, *Statistics of Coal*, pp. 367-368.

engaged in the shipping business.[15] The boatmen, a rough breed, were not averse to selling portions of their coal cargoes along the line of the Schuylkill and pocketing the money. The loss of weight was promptly adjusted by adding quantities of water to the load. This practice eventually was detected, and the Schuylkill Navigation Company insisted that two or more upright tubes be fixed permanently in the bottoms of the boats. A sounding rod lowered into the tubes determined the amount and weight of water, which then was deducted from the total tonnage.[16]

Coal orders were placed for delivery with the shippers or the mine owners by retail yards in Philadelphia and other cities. Some of the large retail dealers such as J. R. and J. M. Bolton, of Philadelphia, maintained offices in Pottsville, bought directly from the mines, and transported the coal in their own boats.[17] The mining of Schuylkill anthracite usually was geared to orders from the eastern markets. Sometimes, however, coal was sent on consignment without fixed prices, and Philadelphia dealers organized to set price levels. The practice was deplored by many coal producers who blamed the high arranged price of anthracite on the retailer.[18] The truth of the matter was that both arranged price levels. Each group, however, contended it was forced to do so to protect itself from the other.[19] On more than one occasion Philadelphians were informed by the *Ledger* that the mine operators, through the Coal Mining Association of Schuylkill County, rigged the wholesale rates of anthracite to suit themselves. The operators also were accused of underproducing in order to create shortages and drive up prices. The *Miners' Journal* vigorously denied these charges and resorted to journalistic "Know-nothingism." The *Ledger* was branded, among other things, a "Jesuitical penny sheet" for daring to suspect collusion among the coalmasters![20]

Over the years the *Ledger* had some reason to be suspicious of price fixing. During the depression in the coal trade in 1849 and 1850, the *Journal* became exceedingly critical of coal operators who, in order to reduce expenses, kept their pumps and hoists in con-

---

[15] Jones, *Anthracite Canals,* pp. 131, 139; CSHS, Palmer Papers, Spruce Street Wharf [ca. 1856].

[16] *Eighth Annual Report Made by the Board of Trade to the Coal Mining Association of Schuylkill County.* Reprinted in *The Pennsylvanian,* February 14, 1840.

[17] HSSC, Eyre-Ashhurst Papers, Coal Receipts.

[18] *Miners' Journal,* May 5, 1849.

[19] *Republican Farmer and Democratic Journal,* March 21, 1849; *American Railroad Journal,* November 16, 1850, p. 742.

[20] *Miners' Journal,* May 2, 1857.

tinuous operation. Continuous operation of mining machinery meant overproduction, groaned the *Journal*. Even in good years this was dangerous. In a depression, overproduction spelled disaster. The trade paper urged unity among individual operators.[21]

During the fifties the vitriolic editor of the *Miners' Journal*, Benjamin Bannan, selected a worthy adversary in Henry C. Carey. One of the richest mines in Schuylkill County was controlled by Carey and his associates through a mining corporation. Bannan, himself a mine operator, for years had fought mining corporations and opposed a general mining law which would eliminate individual appeals for incorporation made to the legislature by mining firms. His greatest fears were overproduction and falling prices, brought about by corporations which, he contended, would flood the market with coal and drive the individual producer into the abyss of bankruptcy and ruin. Bannan, nevertheless, campaigned for a decade in favor of a general manufacturing law for the incorporation of industrial firms, asserting that it would encourage the establishment of factories using anthracite steam power.[22]

Nature favored small enterprise in the Pottsville region more than in the other anthracite districts. Seams of coal lying relatively close to the surface could be worked with small amounts of capital. It was not until after 1865 that many of these seams were exhausted and larger capital investment was required to extend shafts deeper into the earth. Most of Schuylkill County's coal, therefore, was mined by individual operators, who carried on a ceaseless bombardment of coal corporations through legislative petitions and through editorials, letters, and articles in the *Miners' Journal*. For thirty years the paper hurled its epithets at the head of corporative enterprise in the coal industry, bitterly assailing the Lehigh Coal and Navigation Company, the Delaware and Hudson, and similar organizations which invaded the anthracite regions. The Schuylkill Navigation Company, though it possessed no mining privileges, also came under attack and was accused of imposing unfair and monopolistic rates. The Philadelphia and Reading Railroad was welcomed by the mine owners as a competitor to the canal. For a time there was bitter competition between the two modes of transportation, but in the early fifties competition grew into cooperation. It was then the operators attacked the railroad. In a flurry of activity during the winter and spring of 1851, they

---

[21] *Ibid.*, March 13, 1849, April 20, 1850.
[22] *Ibid.*, February 8, 15, 1851.

called for a "Peoples' Road" to market. When it was realized that the initial cost of the railroad would be over five million dollars, the movement fell apart.[23]

It was not until 1870 that the Reading absorbed the Schuylkill Canal and the lateral rail lines which led from the mines to their own tracts. Even then the canal was not to be ignored as an important coal carrier to Philadelphia; in the period before 1860, in spite of railroad growth, the canal fell short of the Reading's coal tonnage by only about half a million a year, as these figures indicate.[24]

Tons of Anthracite Coal Received at Philadelphia*

Carrier	1857	1858	1859	1860
Reading Railroad	1,709,691	1,542,646	1,632,932	1,946,195
Schuylkill Navigation	1,275,988	1,323,804	1,372,109	1,356,678
Delaware Division (Pa. Canal)	530,911	512,512	—	639,323
Others (estimated)				150,000
				4,092,196

* *Annual Reports, Philadelphia and Reading*, 1857-61; *Twenty-sixth Annual Report, Philadelphia Board of Trade*, 1859, p. 146; C. L. Jones, *Anthracite Canals*, tables.

The Reading, after the Civil War, entered the southern anthracite field and gained control of many coal lands. The railroad soon became the largest tonnage transporter in the nation, until surpassed by the Pennsylvania Railroad in 1875.[25] From the beginning, the Reading was a coal railroad, and the tonnage hauled in excess of anthracite was incidental to the overall returns the company received from coal rates. Unlike canal transport, seasonal change had little effect on the railroads. The Reading did not haul as much coal in the winter as in the other seasons due to high maintenance costs in equipment; but it could haul coal, and did, after freezing temperatures had closed the canal routes to Philadelphia.

Other advantages over canal transport became apparent as the company grew with the coal industry. A round trip from the mines to tidewater took less than twenty-four hours. By way of Schuylkill Navigation the round trip amounted to five or six days. Coal cars were shifted from the spur lines out of the mining area to the Reading

---

[23] *Ibid.*, February 22, March 1, 8, 29, April 5, 26, June 7, 1851.

[24] *Annual Reports, Philadelphia and Reading* (1857-1860).

[25] R. W. Brown, "Some Aspects of Early Railroad Transportation in Pennsylvania" (address delivered before the Pennsylvania Historical Association, Dickinson College, Carlisle, October 21, 1949), p. 12.

tracks without delay, waste, or additional expense of unloading and reloading. The railroad prospered. By 1860 the Reading's docks, wharves, and warehouses at Port Richmond on the Delaware, about three miles from central Philadelphia, covered approximately fifty acres. The warehouses could store a quarter of a million tons of coal. Twenty wharves, some eleven hundred feet long, stretched into the river. Ten miles of track wove iron designs back and forth across the area, enabling locomotives to haul the coal cars to the sides of the vessels along the wharves. The bottoms of the cars opened and the holds of the ships were filled quickly. A brig of 150 tons could take on a cargo of coal in less than three hours. At Port Richmond one hundred brigs and schooners could be loaded at the same time.[26] In 1845, three years after the railroad extended to Mount Carbon below Pottsville, it was hauling half a million tons to Port Richmond. Five years later the line transported more than a million and a quarter tons and by 1860 approached the two million-ton mark, or nearly half the total coal tonnage brought to Philadelphia in that year.[27]

Compared with the tonnage shipped over Philadelphia's three major lines of coal transport and their connections—the Lehigh Navigation via the Delaware Division of the State Canal, the Schuylkill Canal, and the Reading—the amount of anthracite shipped over other avenues of commerce to the city was small. In the fifties a few thousand tons came by way of the North Pennsylvania Railroad, which drew its supply from the Lehigh area, while a hundred thousand tons a year came to Philadelphia from the western end of the Wyoming Valley by way of the canal improvements paralleling the Susquehanna and through the Chesapeake and Delaware Canal for transshipment to New York.

The New York market was a primary target of the Pennsylvania coal producers from the beginning of the coal trade. New York, by virtue of her growing population, location, and port facilities, became the greatest single anthracite market in the East. Much of the coal brought to Philadelphia was shipped coastwise to New York by schooner, sloop, and brig. Additional amounts were sent over connecting canals and rail lines which stemmed from Philadelphia. Although Philadelphia functioned as a springboard of supply, other more direct routes were developed from the anthracite regions to New York and its environs.

---

[26] Freedley, *Philadelphia and Its Manufactures,* pp. 116-117.
[27] *Annual Report, Philadelphia and Reading* (1842-1860); Saward, *The Coal Trade,* 1878, p. 6.

Token shipments of Rhode Island anthracite had reached New York during the War of 1812. Small amounts of Wyoming coal which had been floated down the Susquehanna River to the eastern seaboard found a market in the city prior to 1820. New York's first experience with anthracite fuel on a successful commercial basis, however, was provided by the Lehigh Coal and Navigation Company. In 1821 and 1822 the company shipped coal through Philadelphia directly to individual consumers in New York. A year later, a retail yard was established in the latter city. Until 1835 the Lehigh company maintained retail yards in New York; thereafter wholesale contracts were made with individual dealers.[28] By this time the coal trade was established on a firm basis, and the company no longer deemed it necessary or profitable to sell directly to the consumer.

New York and New Jersey businessmen became interested in the future expansion of the coal trade from the Lehigh region not long after shipments of Lehigh anthracite via Philadelphia began arriving in New York and adjoining towns. Many believed large profits could be made on freight if a more direct route were to be established from the mines to the New York market. Consequently, a project was proposed to link the Delaware and Passaic rivers by a canal intersecting the Lehigh Navigation at Easton and running across New Jersey to Newark and Jersey City. A corporation was formed known as the Morris Canal Company, and in 1824 a charter was obtained from New Jersey. The following year the organization opened its books for subscription in New York. The speculative fever of the "canal age" was at its peak. Within one week eager investors had subscribed twenty million dollars' worth of stock. On the strength of this encouraging beginning work was started on the canal. Unfortunately, subscription did not mean cash payment and it was not long before hundreds of subscribers began forfeiting their shares. Apprehension over the forbidding engineering problems of inadequate water supply and the great rise and fall to be overcome by inclined planes, as well as the slow progress of the work, frightened away capital. By 1828 three thousand investors had relinquished their subscribed holdings, and the canal company found itself in a dangerous financial condition. For years it struggled from one crisis to another, nearly expiring in the Panic of 1837.

Like the Union Canal, the narrow locks and shallow water of the Morris Canal admitted boats of only twenty-five-ton capacity and

---

[28] Shegda, "Lehigh Coal and Navigation Company," pp. 169-170.

spelled disaster for a lucrative coal traffic. Poor returns and near-bankruptcy finally led to a thorough reorganization. Supported by state aid from New Jersey and foreign loans, the canal was enlarged. In 1841 boats of fifty-four tons could pass through its locks, but these carriers still could not meet the demands of the increased market and the competition of other routes to New York. On the eve of the Civil War additional improvements increased the canal's capacity to pass seventy-ton boats, yet during the prosperous war years the annual anthracite tonnage never reached the half-million mark.[29] Approximately one-half of this amount was shipped by the Lehigh Coal and Navigation Company and the rest by independent concerns which had been established in the Lehigh coal field in the eighteen-forties and fifties.[30] Large amounts of the anthracite never reached the outlets to the New York market at Newark and Jersey City, but were purchased by anthracite blast furnaces and ironworks which had grown up along the route of the Morris Canal.

The inadequate accommodations of the Morris Canal forced much of the Lehigh coal trade to descend to New York over the longer route of the Delaware and Raritan Canal, which had been constructed in the mid-thirties. This canal joined the Delaware River at Bordentown, ran northeast to Trenton, crossed New Jersey, and terminated at New Brunswick. The coal boats then were floated down the Raritan River to Perth Amboy and the New York area. The vital link in the route was the Delaware Division of the Pennsylvania State Canal. In order to gain a return on the coal traffic to Bristol, Pennsylvania made the Lehigh coal boats bound for the Delaware and Raritan return up the river to Bordentown. The coal interests persisted in demanding an outlet lock near Trenton which would eliminate the additional expense and time of the trip to and from Bristol, but the legislature blocked their appeals for a number of years. Finally, when the improvements on the Morris Canal threatened to appropriate additional anthracite tonnage, the Pennsylvania Canal Commission recommended the construction of an outlet lock at Wells Falls above Trenton.[31] Although the route was shortened by twenty-six miles, the State of Pennsylvania pursued a narrow policy and collected the same amount of toll whether the coal descended to Bristol or moved to the Delaware and Raritan Canal by way of Wells Falls.

---

[29] Jones, *Anthracite Canals,* pp. 104-117.

[30] *House Executive Document,* No. 65, 32nd Congress, 2nd Session, *Report of the Commissioner of Patents,* 1852, p. 424.

[31] *Pennsylvania Senate Journal,* 1846, II, 257-259.

In 1857, two weeks before the Delaware Division was sold to railroad interests, the Commonwealth, in an empty gesture, agreed to reduce charges to meet increased railroad competition.[32]

The railroads provided additional transportation necessary to supply the expanding anthracite markets. During the fifties large amounts of Lehigh coal began moving by rail to the New York area. The Lehigh Valley Railroad, which began operations in 1855, served as a rail trunk from Mauch Chunk to Easton, where it connected with two branch lines. One, the Belvidere and Delaware, followed the Delaware Valley south to Trenton and transferred the coal to the canal boats on the Delaware and Raritan. The other branch line, the New Jersey Central, became famous as an anthracite coal carrier, drawing its tonnage from not only the Lehigh Valley Railroad but also the Delaware, Lackawanna and Western out of Scranton in the Wyoming fields. The New Jersey Central terminated at Elizabethport on New York Bay, thirteen miles southwest of New York City.[33]

The new rail outlets from the Lehigh and Wyoming fields disturbed the officials of the Reading Railroad, for at least one quarter of their Schuylkill coal deposited at Port Richmond went by sea or canal to the New York market.[34] Between 1855 and 1860 the coal tonnage hauled by the Reading dropped noticeably and did not surpass the peak of 1855 until the second year of the Civil War. In 1857 the company blamed the panic for its losses, but neglected to inform its stockholders that during and after this crisis anthracite tonnage had increased on the Lehigh Valley and Delaware, Lackawanna and Western lines.[35]

Although the Reading experienced a decrease in trade within this five-year period due chiefly to new avenues of supply to New York and vicinity, the broad picture indicates that the New York market continued to expand. Schuylkill anthracite, an early arrival in the city, maintained its primary position there. Supplies forwarded by the Reading were bolstered by the older route, the Schuylkill Canal. Independent operators in the Pottsville area instituted a new policy in 1839 of shipping some coal "direct" to New York in covered boats.

---

[32] Jones, *Anthracite Canals,* pp. 66-67.

[33] Saward, *The Coal Trade,* 1877, p. 5; J. D. Steele, *Letter to the President of the Philadelphia and Reading Railroad Company on the Canals and Railroads for Transporting Anthracite Coal, 1855* (Philadelphia, 1855), p. 5.

[34] Steele, *Letter . . . on . . . Transporting Anthracite Coal,* p. 17.

[35] *Annual Report, Philadelphia and Reading* (1858-1861).

The boats moved down the canal to Philadelphia where they were towed up the Delaware to the intersection of the Delaware and Raritan. This procedure saved time, expense, and waste in unloading, storing, and reloading at Philadelphia.[36] By the mid-fifties Schuylkill Navigation was sending over a half-million tons per year to the New York area.[37] This amount, when added to the Reading shipments out of Port Richmond, accounted for more than half the anthracite tonnage received at New York. The total from all sources of supply averaged two million tons annually in the few years preceding the Civil War.[38]

Perhaps the most interesting coal firm to supply the New York area was the Delaware and Hudson Canal Company, which mined and carried Lackawanna anthracite of the eastern Wyoming field to the Hudson River and tidewater. The enterprise had its beginnings in the activities of Maurice Wurtz and his brothers, Charles and William, all of Philadelphia. The chief object of the Wurtz brothers was the mining and transporting of coal to New York City and Hudson River towns. Gaining charters from the states of Pennsylvania and New York in 1823, the company issued stock and began work in 1826. At first a rough wagon road, then later a combination gravity and inclined-plane rail line ran from the mines at Carbondale over a ridge of mountains to a wilderness threaded by the Lackawaxen Creek. The company erected a village on the creek and named it Honesdale after Philip Hone, the first president of the Delaware and Hudson. Honesdale became the starting point of the canal. Paralleling the Lackawaxen to where it joined the Delaware River, the canal followed the eastern path of the Delaware Valley for about twenty miles to Port Jervis and then struck northeast across New York for sixty-seven miles to Rondout on the Hudson River.[39]

Rondout was located strategically for the prospective markets of the company, since it was sixty miles below Albany and some ninety miles above New York. From Rondout the coal boats moved north and south along the Hudson, supplying towns from the gate of the Mohawk Valley to the river's mouth. Towed north to Albany, some of the boats continued through the Erie Canal west to Schenectady,

[36] Hazard, *U.S. Register*, II, January 22, 1840, p. 59.

[37] *Annual Report of the Schuylkill Navigation Company* (1855), p. 25.

[38] *Annual Report of the Chamber of Commerce of the State of New York. For the Year 1858*, p. 84.

[39] Hazard, *U.S. Register*, V, September 15, 1841, p. 164.

Utica, Syracuse, Rochester, Buffalo, and the Great Lakes market. Other boats moved south to New York and transferred their loads to brigs and schooners which sailed Long Island Sound to Connecticut towns or proceeded through the narrow, rock-strewn entrance of Narragansett Bay bound for Providence. A few joined the growing fleet of colliers out of Philadelphia and with them beat their way around Cape Cod to Boston. Both the western New York-Great Lakes trade and the New England market had unpretentious beginnings, but were to prosper with the years as the homes and industries of these regions demanded growing quantities of anthracite coal.

The Delaware and Hudson was one of the most progressive companies in the anthracite trade. Conscious of the changing American economic scene, the coal and canal firm promoted the product of its mines through agents in Albany, Providence, Boston, and New York. A few hogsheads of Lackawanna anthracite were even forwarded to New Orleans for demonstration purposes. The largest and most important market, however, remained New York. Lackawanna coal, considered the lightest of the white-ash variety, possessed a lower proportion of fixed carbon than the Lehigh or Schuylkill anthracite. It ignited more readily and burned with a quick, high heat. The coal was excellent for steam purposes and was used extensively by factories and steamboats. The introduction of anthracite coal as steamboat fuel was due primarily to the efforts of the Delaware and Hudson Canal Company, which saw in this new application a potential market of large consequence.[40]

The first boats to pass over the canal in 1829 were small, thirty-ton craft. Twenty years later improvements permitted boats of eighty tons to move through the Delaware and Hudson's locks. The coal tonnage increased, with enlarged transport facilities and growing demand, from seven thousand tons in 1829 to one hundred and fifty thousand in 1840. Within six years the 1840 tonnage had doubled because of steamboat consumption. In 1850 the company mined and sent to market nearly half a million tons. In that year the Pennsylvania Coal Company at Pittston began shipments to New York via rail connections from its mines to the Delaware and Hudson Canal. Within four years Pittston coal surpassed the tonnage of the older firm, contributed to the large increase in tolls over the canal route, and furnished anthracite for the same channels of trade supplied by

[40] Delaware and Hudson Minute Books, 1829-1850.

the Delaware and Hudson.[41] The fact that these markets were able to absorb twice as much anthracite from the same line of supply between 1850 and 1854 is clear evidence of their ability to expand without disastrous economic consequence.

For many years anthracite coal had moved north by way of the Hudson and west along the line of the Erie Canal to the inland towns of New York and the Great Lakes. Lehigh and Schuylkill coals could be added to the Lackawanna anthracite which found ready sale in these regions. Before 1850 the western anthracite trade was never a large one, but one, nevertheless, which grew steadily and had great possibilities. The State of Pennsylvania was aware of the potential. The state of New York had no coal, a fact repeated again and again by the voices of Pennsylvania economic sectionalism. Coal was to be the bait used to lure the western commerce from the Great Lakes and the Erie Canal south to Pennsylvania. All that was needed was an adequate canal connection from the northern anthracite fields to western New York. Anthracite coal then would move north to be exchanged for salt, gypsum, grain, and other commodities. The line of the Erie could be tapped and the profits of the western trade would accrue to Pennsylvania.[42]

In 1834 the Packer Committee, reporting on the coal trade to the Senate of Pennsylvania, estimated the western New York-Great Lakes market to be one hundred thousand tons, the majority of which was sent by way of the Hudson River and Erie Canal. During the winter small amounts of anthracite also were hauled by sled into western New York, bringing as much as twenty dollars per ton. If Pennsylvania would construct a canal from the anthracite fields to connect with the New York State improvements, announced the report, the tonnage could be tripled.[43] A canal along the North Branch of the Susquehanna River was proposed and received the support of Governor Joseph Ritner and members of the legislature. A "vast and profitable business" in anthracite was to be developed between the northern coal fields and Elmira, Binghamton, Ithaca, Oswego, and the "Lake Country."[44]

The early promise of a lucrative trade from the northern field into the state of New York was not fulfilled by virtue of the North Branch

---

[41] *Delaware and Hudson Company Annual Report* (1864), p. 18; Taylor, *Statistics of Coal*, p. 384; Jones, *Anthracite Canals*, pp. 74-85.

[42] *Pennsylvania Archives,* Fourth Series, V (1831), 993, 998.

[43] *Report to the Senate of Pennsylvania on the Coal Trade*, p. 29.

[44] *Pennsylvania Archives*, Fourth Series, VI (1837), 316-318, 380.

Canal. Pennsylvania plunged deeply into debt building her canal lanes and eventually abandoned them to private interests.[45] Then, too, the development of the coal fields around Wilkes-Barre and Scranton was hampered by lack of capital in the thirties and forties. Most of the coal mined in those areas sought outlet to the south by way of the Susquehanna River improvements. Until the decade of the fifties anthracite tonnage between Pennsylvania and New York over the more direct connection of the North Branch Canal lagged far behind the Hudson-Erie route. In the fifties railroads were extended from the northern anthracite regions into western New York and the Great Lakes area, intersecting New York rail lines. The Sunbury and Erie, the Catawissa, Williamsport and Elmira, the Lackawanna and Western from Scranton to Great Bend, the Lehigh and Susquehanna, the North Pennsylvania Railroad out of the Lehigh and Wyoming fields to Waverly, New York, the New York Central, and the New York and Erie forged an iron lattice of transport over which were hauled increasing tonnages of Pennsylvania anthracite coal.[46]

The North Branch Canal of the Susquehanna River functioned in a dual capacity. First, it served as a channel of trade to the northern markets; second, it flowed southwest to join the Susquehanna Division of the Pennsylvania Canal near Sunbury, which, in turn, connected with the Susquehanna and Tidewater Canal for the remaining length of the river until it emptied in Chesapeake Bay. Boatloads of anthracite moved down these improvements to Port Deposit and Havre de Grace, where larger vessels or steam tows were waiting to take the coal to Philadelphia, Baltimore, Washington, and smaller coastal or river towns.

Baltimore became the most significant anthracite market south of Philadelphia. The extraordinary development of the Baltimore anthracite market in the face of competition from the cheaper bituminous Cumberland coal indicated the advantages of the Pennsylvania product, particularly for domestic use. Cumberland coal reached the Maryland port by way of the Baltimore and Ohio Railroad and

---

[45] *Ibid.*, VII (1857), 934-937.

[46] For a general account see J. I. Bogen, *The Anthracite Railroads* (New York, 1927); see also *Pennsylvania Senate Journal*, Appendix to II, 1841, pp. 10, 41, 45; *Pennsylvania Legislative Documents*, 1854, pp. 416-418; *Pennsylvania Our Internal Improvements* (n.d.); *Merchants' Magazine*, XXXIII (July-December, 1855), 323-329; *First Annual Report of the Board of Directors of the North Pennsylvania Railroad Company*, January 9, 1854 (Philadelphia, 1854), pp. 4-15.

the Chesapeake and Ohio Canal. Pennsylvania anthracite was shipped down the Susquehanna to the Chesapeake and later over the Northern Central Railroad, which drew much of its coal from the vicinity of Mount Carmel in the Shamokin Basin. From 1850 to 1860 the amount of anthracite entering the Baltimore market averaged a quarter of a million tons a year.[47] By 1860 the value of Baltimore anthracite commerce was set at $1,600,000 as compared with $1,500,000 for her Cumberland bituminous trade. The competitive effects of anthracite at Baltimore and in other eastern cities, coupled with duty-free Nova Scotia bituminous "flooding" northern markets, caused some observers to lament that the bituminous export trade of Baltimore would soon become a thing of the past.[48]

Baltimore was the important southern terminus of the Pennsylvania anthracite trade, while Boston represented the northernmost significant market. The Lehigh Coal and Navigation Company was the first concern to sell to the New England area, concentrating at Boston but establishing small retail yards at Hartford, Providence, and Portland in 1824 and 1825. A few years later Schuylkill coal was shipped to New England towns, with Boston as the major port of entry. In the beginning the market was small. Importing firms, such as Jonathan Amory and Son, of India Street, Boston, which sold numerous items from Canton window blinds and table mats to choice Madeira wines, from time to time advertised Schuylkill coal for sale by the cargo.[49] Retail dealers bought from Boston importers or ordered directly from Philadelphia shippers who sold both wholesale and retail on the Boston market.[50] The North American Coal Company, one of the few early corporations in Schuylkill County, offered its Peach Orchard anthracite to Bostonians through an agent who received orders at Philadelphia "for shipping at short notice."

The Delaware and Hudson also sought the New England area as an outlet for its Lackawanna coal. Like the Lehigh company, the Delaware and Hudson sent agents to New England to introduce the new fuel. Lackawanna anthracite arrived in Boston for the first time in the spring of 1830 and sold for six dollars a ton plus shipping charges from Boston. An item in the Boston Daily Advertiser advised, "Persons unacquainted with this coal will be furnished with samples for trial." Orders were taken either through the company

---

[47] Miners' Journal, January 15, 1859, January 14, 1860.
[48] Merchants' Magazine, XLII (January-June, 1860), 570-571.
[49] Boston Daily Advertiser, September 11, 16, November 16, 1829.
[50] Ibid., February 24, 29, March 17, October 17, 26, 1829, February 12, 1830.

agent, a Mr. N. T. Eldridge, or William H. Prentice and Son, importers. The coal was equal to any anthracite on the market, boasted the advertisement, and purchasers were invited to examine the fuel at the Prentice wharf.[51]

Through the decade of the thirties, anthracite, although competing with bituminous coal from Liverpool, Nova Scotia, Maryland, and Virginia, gained a foothold on the Boston market. In 1845 one hundred and sixty-five thousand tons were received by the port. Half the amount was used by industries and the other half for domestic purposes.[52] By 1847 coal imports exceeded a quarter of a million tons, and Boston was looked upon by Pennsylvania anthracite interests as an exceptionally promising market. Inhabitants of Boston and other New England cities and towns accepted the fuel with growing confidence. Anthracite steam-powered factories and anthracite stoves and grates for homes and public buildings multiplied in the northern region. Between 1850 and the Civil War, Boston's anthracite tonnage averaged nearly four hundred thousand tons per year: a remarkable growth when one considers the retail methods of the early coal dealers and the great distance from the sources of production.[53]

The attempt to establish a European market for Pennsylvania anthracite coal was made by Friedrich List, German economist and political refugee from the kingdom of Württemberg. List found a haven in Pennsylvania, first at Harrisburg and later at Reading where he became a citizen of the United States and edited a German-language newspaper, the *Readinger Adler (Reading Eagle)*. While on an expedition to the regions of the upper Schuylkill Valley, he chanced upon an outcropping of anthracite coal along the Little Schuylkill River at the present site of Tamaqua. Aided by his friend, Dr. Isaac Hiester, a prominent Reading physician, List prevailed upon Philadelphia capitalists, including Stephen Girard, for funds to organize a mining and transportation company known as the Little Schuylkill.[54]

In 1830, a year before the Little Schuylkill Railroad began drawing coal cars by mule and horse over its newly laid track. List journeyed to Europe.[55] His departure was hastened by personal commercial

---

[51] *Ibid.*, May 21, 22, 1830.

[52] Taylor, *Statistics of Coal*, pp. 458-459.

[53] See Table H, Appendix.

[54] M. E. Hirst, *Life of Friedrich List and Selections from his Writings* (New York, 1909), pp. 55-63.

[55] It was not until 1833 that the English steam locomotives, the "Comet" and the "Catawissa," arrived in the United States for service on the Little Schuylkill Railroad.

motives. List's notes explaining the purpose of the trip indicate that he was interested primarily in establishing a market for Little Schuyl-kill anthracite in France and the Germanies.[56] During the few years the economist had been in the United States, he had made the ac-quaintance of numerous important persons, including two key men in the Jackson administration, Martin Van Buren and Edward Living-ston. Through these connections he was able to secure a commission as American consul at Hamburg. The position of consul would serve as a means to an end. What better way was there to promote his plans than as an official commercial representative of the United States? List even offered to serve without remuneration, stipulating that the incumbent, John Cuthbert, who had a large family and no other means of support than the consular post, continue to be paid his salary. His letters to Van Buren, William C. Rives, American minister to Paris, and to Cuthbert in Hamburg have all the flair and enthusiasm of a traveling coal salesman. He probed for informa-tion concerning tariff agreements, costs of wood and British coal, types of stoves, and port facilities. List reported to Van Buren, in glowing terms, the prospects of a lucrative European market in Pennsylvania anthracite coal.[57]

List's hopes were dashed by the cool reception the French gave to his proposition of a commercial treaty which he made to the "July Monarchy" through Rives. At this time the question of the claims for spoliation under Napoleon's Berlin and Milan decrees was before Louis Philippe. Rives wrote to List that "so long as this horrible question of the claims remains unsettled," there was little chance for a trade agreement. The new French government feared adverse pub-lic opinion, which would believe that France was paying for the claims through commercial concessions.[58]

Even with the duty of ten francs per ton, List felt there could be a market if the French were educated in the advantages of anthracite home heating and cooking. A close friend of General Lafayette, List probably was the one who proposed that samples of anthracite and an iron stove be sent "to the great benefactor of the human race"

---

[56] E. V. Beckerath, Karl Goeser, *et al.* (eds.) , *Friedrich List, Werke* (Berlin, 1928-1933) , II, 483-484.

[57] *Ibid.*, Friedrich List to John Cuthbert, January 10, 1831, p. 496; Cuthbert to List, February 1, July 12, 1831, p. 497; List to Martin Van Buren, January 7, 1831, p. 305; List to William C. Rives, November 15, 1830, p. 492.

[58] *Ibid.*, Rives to List, December 30, 1830, pp. 492-493.

whose prestige and experiments, reported a Philadelphia newspaper, undoubtedly would give Pennsylvania the coveted trade.[59] The French did not adopt the strange fuel. List then focused his attention on the Germanies. Here he met with prejudice similar to that of the early days of the trade in the United States. Even more discouraging, as well as embarrassing, the United States Senate refused to confirm his nomination as consul at Hamburg. This was due to the long and faithful service of Cuthbert, and to the protests of German reactionaries who insisted that List, a "dangerous" political radical, would not be welcome.

List returned to Pennsylvania late in 1831, gained an appointment as United States consul to Leipzig, and went back to Europe to fill his position in the face of conservative opposition. He served at Leipzig, and at Stuttgart in his native Württemberg, which must have received him with some reservations, although he had gained political respectability as an authority on a German railway system. His interest in anthracite coal was now confined to the income he received from his original investment in the Little Schuylkill. List did not attempt to promote a European market for Pennsylvania anthracite after his failure in 1830-31. Instead, he concentrated on the construction of a German railway system with all the energy his brilliant, visionary mind could muster. But pressing financial affairs, official duties, and grandiose schemes, only half-fulfilled, for a Germany commercially unified by the iron rails of transport led to a physical and mental breakdown. He died by his own hand in the snows of the Tyrolean Alps in November, 1846.[60]

After the failure of Friedrich List to introduce Pennsylvania anthracite to continental Europe, no other efforts were exerted by the coal interests throughout the remaining thirty years of the period. From time to time, a few tons of anthracite would be shipped across the Atlantic, usually as ballast, but the amount always was trifling.[61] Sometimes a cargo would find its way to England. Yet so unusual was this that when it did occur men with long experience in the trade expressed surprise and amusement.[62] For the most part anthracite

[59] *Poulson's American Daily Advertiser*, January 4, 1831.

[60] R. W. Brown, *Friedrich List—the Father of German Railroads—His Residence in Dauphin and Schuylkill Counties, Pennsylvania* (address delivered before the Historical Society of Dauphin County, Harrisburg, September 18, 1950), pp. 8-9.

[61] *Twenty-seventh Annual Report of the Philadelphia Board of Trade* (1860), p. 121.

[62] Nevins, *Diary of Philip Hone*, II, 678.

producers concentrated on the domestic trade, both inland and coastal, focusing their attention on the development of the home market.

The anthracite trade, however, did extend to Canada by sea and via the Great Lakes. Pennsylvania coalmasters looked upon the Canadian Reciprocity Treaty of 1854 with favor after it had been in effect a year or two, and staunchly supported it against Philadelphia and New York interests who were dissatisfied with the measure.[63]

As the coal trade extended north to Canada, it also moved south to Florida, Alabama, Louisiana, and Texas. By 1860 a thousand tons or so a year reached California by way of the Isthmus of Panama or around Cape Horn. After the Treaty of Whanghia in 1842, China, familiar with coal for a thousand years, permitted Pennsylvania anthracite to enter her ports. Philadelphia, Boston, and New York shippers exported an average ten thousand tons per year to the Far East in the late fifties. On the other side of the world, British African possessions imported a few thousand tons of Pennsylvania anthracite yearly.[64]

Latin American countries and the islands of the West Indies furnished another foreign market for anthracite coal. Brazil, Chile, Peru, New Granada, Mexico, and the Caribbean islands purchased a combined amount of over ten thousand tons in 1859. Before the Civil War there was hardly a port in Latin America which had not unloaded a cargo of Pennsylvania anthracite coal at one time or another.[65]

The export of anthracite, though never very large—ranging from fifty to a hundred thousand tons a year—nevertheless spread over much of the world. From Hawaii, the Philippines, the China coast, and the British East Indies to Africa and the countries of Latin America, Pennsylvania anthracite coal became, if not a plentiful, at least a familiar product of import.

THE WESTERN COAL TRADE

The bituminous coal trade of western Pennsylvania before 1860 did not develop as rapidly or on such a large scale as the anthracite commerce of the eastern part of the State. For many years western coal

---

[63] *House Executive Document,* No. 96, 36th Congress, 1st Session, 1860, p. 60.

[64] See *Annual Report of the Philadelphia Board of Trade* (1858), p. 67; (1859), p. 121; *Annual Report of the Chamber of Commerce of the State of New York. For the Year 1859,* p. 87.

[65] *Ibid.*

was used locally by inhabitants along the Monongahela and Youghiogheny rivers and by the people of Pittsburgh. The sparse population of the western country, undeveloped transportation facilities, and adequate supplies of timber prohibited the extensive growth of coal markets for a generation after the nineteenth century had begun.

Forced to turn west and south by the great ridges of the Alleghenies, settlers on the frontier found no other paths of trade save those of the rivers which converge at Pittsburgh to form the Ohio. From Pittsburgh the mighty Ohio flows north, then dips south and west for nearly a thousand miles to join the Mississippi. It was the river road, made up of the Monongahela, the Youghiogheny, the Ohio, and the Mississippi, which formed the main artery of the Pennsylvania bituminous coal trade. With the increase of population in the Ohio and Mississippi valleys, the growth of cities, and the clearing of forests, fuel needs, for the most part, were supplied by Pennsylvania bituminous coal mined in the vicinity of Pittsburgh or along the Monongahela and Youghiogheny rivers.

Just a few years after the end of the War of 1812, a small coal traffic began to move from Pittsburgh down the Ohio. Some of the coal was mined in the vicinity of the town itself and some along the Monongahela River. In the Monongahela Valley, coal was mined during the winter and wheeled in barrows or in carts drawn by dogs or mules to "stockyards" along the banks of the river. With the spring rise, the coal was loaded into flatboats or "French Creek" boats. These latter craft were large flats which had been used to float iron ore from the French Creek region down the Allegheny to Pittsburgh. After being unloaded, they were sold to coal men who poled them up the Monongahela to their mines. The boats then were "sided up" to increase their capacity to about six thousand bushels or two hundred and forty short tons. The boats drew considerable water when loaded and the Monongahela had to be nearly at flood stage before they could be floated to market, either singly or lashed together in pairs.[66]

From Pittsburgh, with luck and good water, it was a five-day journey to Cincinnati where, by the eighteen-twenties, there was to be found an increasing market for bituminous coal. By 1840 Cincinnati used an estimated two and a half million bushels, the greater part of which came from Pennsylvania. In less than twenty years the amount

---

[66] Wall, *Second Geological Survey of Pennsylvania, Report of Progress K4*, pp. xxiv-xxvi.

had risen to fifteen million bushels, while the year the Civil War began the city received seventeen and a half million bushels or approximately seven hundred thousand tons.[67]

Pittsburgh, however, remained the chief market for both her own mines and those of the Monongahela and Youghiogheny valleys. In 1837 the mines of Coal Hill overlooking the city produced more than five million bushels. Two and a half million bushels were sent down the Ohio. Yet Pittsburgh's estimated consumption for that year was nine and a half million bushels of coal, an amount which was supplemented by the mines of the great, unbroken coal seam which stretched from McKeesport to Brownsville.[68] By 1855 Pittsburgh was using over thirty-three million bushels annually, which was equal to well over a million tons. Besides this, the wharves of the city re-exported from the Monongahela district half a million tons, and another quarter of a million tons from the Youghiogheny Valley.[69]

In the fifties Pennsylvania bituminous coal was in competition with western Virginia, Ohio, and Kentucky coals in the Ohio and Mississippi valleys. Still, Pennsylvania production figures climbed and the markets along the Ohio and Mississippi continued to expand. Cincinnati, Louisville, Memphis, St. Louis, Natchez, and New Orleans used growing amounts of bituminous coal in factory and home. The cities became dependent upon fuel supplies from Pennsylvania. The figures quoted in Louisville papers, though incomplete, show an increase in coal received from all sources from thirty-two hundred net tons in 1841 to nearly a quarter of a million tons by 1860.[70] More complete records were kept at New Orleans. Here, coal was sold by the barrel of two hundred pounds, or two and a half eighty-pound bushels to the barrel, and ten barrels to a net or short ton of two thousand pounds. In 1835 New Orleans received less than five thousand tons of what was listed as "Pittsburgh Coal." Ten years later, the amount had increased to fourteen thousand tons. Figures fluctuated to 1860, but there was steady increase over the years. The amount had risen to over twenty thousand tons in 1850. Ten years later, just before the blockade and the river war closed her port and suspended normal peacetime business, over one hundred and sixty-eight thousand tons of Pennsylvania bituminous coal were unloaded at her

[67] See Table J, Appendix; *Cist's Weekly Advertiser,* February 14, 1848.
[68] Wall, *Second Geological Survey of Pennsylvania, Report of Progress K4,* p. xxvii.
[69] See Table K, Appendix.
[70] See Table M, Appendix.

wharves for home consumption or for transshipment to foreign ports.[71]

Tonnage statistics are insufficient to tell the complete story of the Pennsylvania bituminous coal trade and its expanding river markets. Industrial and domestic utilization of the western coals of Pennsylvania was basic to the growth of the trade. Technological improvements in gasworks; in manufacturing plants where steam power was employed; and in the iron, salt, and glass industries where not only steam power, but great amounts of heat were needed all called for larger quantities of mineral fuel. A growing population, capital investment, and improved transportation facilities also played important parts in the development of the western coal traffic.

Transportation, certainly, was one of the most significant threads in the economic fabric of the coal trade. The hazards of river transportation were many, but despite dangers the coal traffic persisted and prospered. Coal fleets leaving Pittsburgh in the swirling waters of a spring freshet or a winter rise sometimes faced the possibility of being swamped, driven aground, or sunk by floating ice. In seasons of low water, if the coal boats could be floated at all, there were shifting sand bars, rocks, snags, and driftwood islands to tax the alertness of the most skilled river pilots. The people of Pittsburgh and the Ohio River cities watched the waters of the Ohio and the weather with anxiety and concern. "The Weather and Business" and "The Weather and the River" were common newspaper headlines. Often, divine assistance was sought to regulate weather conditions.[72] Too much precipitation or not enough, quick freeze or sudden thaw were equal enemies. The Ohio at Pittsburgh often rose or fell five to seven feet in one or two days. All trade, and particularly the heavy, bulky coal commerce took advantage of the freshets and moved out to mid-stream in an almost frantic effort to make down-river markets.[73] The fall and winter of 1856 were particularly bad seasons of drought. Pittsburgh's population prayed for snow and rain.[74] When a rise came and a coal fleet set out, the rushing waters of the river strewed fifty barges and flats "in a line of wreck" between Pittsburgh and Beaver.[75]

---

[71] See Table L, Appendix.

[72] L. C. Hunter, "Seasonal Aspects of Industry and Commerce Before the Age of Big Business," *Smith College Studies in History*, XIX (October, 1933-July, 1934), 18-19.

[73] *Pittsburgh Gazette*, January through February, 1854.

[74] *Ibid.*, December 2, 4, 1856.

[75] *Ibid.*, December 9, 1856.

Bitter weather followed, closing river traffic for hundreds of miles. In the Ohio cities, the economic plight and the human suffering for lack of fuel were acute. Some relief was brought by new rail connections, but it was not until the river again was ready for transportation that business resumed in a normal manner.

Pittsburghers did little before 1860 except to protest the lack of Ohio River navigation improvements. On the other hand, constructive work was undertaken between 1836 and 1845 to improve an important artery of commerce flowing into their city. The Monongahela River and its principal tributary, the Youghiogheny, formed the major water lanes from the bituminous coal fields to the markets of Pittsburgh and beyond. The Monongahela, unlike the Allegheny River to the north, presented serious navigational hazards. Flatboats loaded with coal at various points along the river were forced to stay loaded and idle, sometimes for months, before there was enough water to float them to market. A skeleton crew had to be retained to watch and bail the boats. During the drought of 1837 the loss to Monongahela coal operators from a serious delay of this kind was set at $40,000 for the wages of the men alone. In another recorded incident in the fall of 1839, 150 flats, each holding five thousand bushels of coal, had been tied up along the Monongahela for three months for lack of sufficient water. It took a crew of five to man each boat when moving downstream. But two men per boat were required to watch and bail each flat tied to the wharves along the river's banks. With coal selling at five cents a bushel at Pittsburgh and double to triple that down the Ohio, and the wages of three hundred men at one dollar a day, the loss could be reckoned at not less than sixty-five thousand dollars.[76]

In 1837 work was begun on the slackwater improvement of the Monongahela River. A series of financial setbacks delayed the completion of the work from Pittsburgh to Brownsville, a distance of fifty-six miles, until 1844-45. When the slackwater canal was opened for navigation the agricultural trade of the Monongahela Valley expanded. But above all coal became the chief item of transport. The value of coal lands along the river increased, new mines were opened, and coal was produced in larger quantities than ever before.[77] The effects of the Monongahela slackwater improvement were felt not

---

[76] *Second Annual Report of the President and Managers of the Monongahela Navigation Company,* February, 1840. Reprinted in the *Pennsylvania House Journal,* 1839-40, II, Pt. II, pp. 38-39.

[77] *Seventh Annual Report, Monongahela Navigation Company,* 1847. Reprinted in the *Pennsylvania Senate Journal,* 1847, II, 375-379.

only in the valley itself, but north to Pittsburgh and down the Ohio and Mississippi as far as New Orleans. In 1845 over four and a half million bushels of coal were mined and shipped to market over the Monongahela Navigation Company's improvement. This was not the total amount to come out of the valley, for there were some coal operators who refused stubbornly to use the new locks. In two years the amount passing over the canal had doubled. By 1852 the Youghiogheny had been slackwatered and its improvements contributed to the growing amount of coal which arrived at Pittsburgh or moved down the Ohio and Mississippi. Production figures continued to mount: in 1855 twenty-two and a quarter million bushels, and in 1860 thirty-eight million or a million and a half tons.[78] Although the Monongahela Navigation Company did not "create" the coal trade, as it stated in reports, its improvements were responsible for the great increase in the western coal traffic.

In 1845 a significant change began in the mode of coal transport on the western rivers. Small steam towboats made their appearance on the Monongahela and flats were towed instead of floated to Pittsburgh. From Pittsburgh the process was repeated to downriver markets. The first steamboat to make this kind of trip from Pittsburgh to Cincinnati was the small stern-wheeler, the *Walter Forward,* piloted by Captain Daniel Bushnell. By 1850 a new method of steam transport was in vogue. The coal flats or barges were placed at the bow of the steamboat and strung out in echelon on either side. The barges were then pushed by the steamer. This arrangement made for greater maneuverability of the barges and also cheapened the cost of transportation by reducing the number of crew to man the barges.[79] As late as 1874, when Pittsburgh was shipping fifty million bushels a year to river markets, this method of transport was more popular than ever. The coal fleets came out on floods of seven feet or more and took advantage, not only of the depth, but of the breadth of the waters of the Ohio and Mississippi rivers. It was not uncommon for one steamer and her barges to spread the width of 125 feet or more.[80]

After 1845 increasing amounts of Pennsylvania bituminous coal were sent to the Great Lakes. By that time a water route extended

---

[78] See Table N, Appendix.

[79] Wall, *Second Geological Survey of Pennsylvania, Report of Progress K4,* p. xxxii.

[80] F. R. Brunot, "Improvements of the Ohio River," *Journal of the Franklin Institute,* LXVII (January-June, 1874), 5.

north from Beaver on the Ohio River to New Castle, where it joined the Erie Extension Canal. By 1855 one hundred and forty thousand tons a year, most of which came from the Shenango mines of Mercer County, were shipped over the canals to Erie, Pennsylvania. Erie became one of the two chief ports of distribution for the bituminous coal trade of the Great Lakes, sending vessels north to Canadian ports and west as far as Chicago and Milwaukee.[81]

The other significant port was Cleveland. Until 1845, when it was possible for Pennsylvania bituminous coal to move north in quantity to Erie, the soft-coal trade of the Great Lakes was in the hands of Cleveland merchants who received both Pennsylvania and Ohio coals. Pittsburgh and Monongahela mines shipped to Cleveland over the Pennsylvania and Ohio Canal by way of Akron. Later, the Pittsburgh and Cleveland Railroad carried additional amounts to the lake port. Ohio coal was brought to the city over the Ohio Canal and its feeders as well as by rail transport from the Ohio coal fields.[82]

Water transport continued to play an important part in the western coal trade for many years after the Civil War. But in the decade of the fifties, railroads already were supplementing the western water routes. Above all, railroads provided new outlets from the bituminous coal fields eastward to the Atlantic seaboard. The amount of western bituminous coal shipped to eastern markets before 1850 was decidedly small. There had been high hopes for a lucrative coal traffic in the balmy days of State improvements. Moncure Robinson, engineer for the Allegheny Portage Railroad, believed that the total tonnage would be much greater from west to east because of the bituminous coal trade. For that reason, he urged the installation of more-powerful stationary steam engines for the planes west of the summit in order to haul the heavy canal boats over the mountains from Johnstown to Hollidaysburg.[83]

Over the years comparatively few coal barges passed that way. Most of the bituminous coal of the State followed the great rivers west and south. Excessive freight rates and narrow locks on the Pennsylvania System, coupled with competition of Maryland, Virginia, and foreign bituminous coals, made the eastern trade unprofitable. Even the

[81] *Merchants' Magazine*, XXXIV (July-December, 1856), 633-634.

[82] *Pennsylvania Senate Journal*, 1845, II, 129-133; Saward, *The Coal Trade*, 1878, p. 52; Eavenson, *American Coal Industry*, Map No. 12; *The Pennsylvanian*, May 21, 1846.

[83] *Pennsylvania Senate Journal*, 1829-30, II, 245.

demands of eastern gasworks for the excellent gas coals of western Pennsylvania were met sporadically until the mid-fifties. As late as 1856 only a few thousand tons moved east over the Portage Railroad and canals.[84] The Pennsylvania bituminous coal which did find its way to eastern markets was mined on the eastern edges of the Great Allegheny Coal Field and was brought to market over the West Branch Canal of the Susquehanna River.[85]

It was not until the Pennsylvania Railroad penetrated western Pennsylvania and joined Philadelphia with Pittsburgh in 1852 that the eastern market assumed significance. Even the more profitable freight of the Pennsylvania Railroad out of Pittsburgh depended on the weather and the river. When the waters of the Ohio were high, generally in the spring and autumn, products of the western country literally flooded into Pittsburgh and provided the Pennsylvania Railroad with goods to be shipped east. Bituminous coal in the Pittsburgh area, or at the Westmoreland mines east of McKeesport, could be obtained as freight in any season. But coal freight brought only $4.70 a short ton, whereas the charges on one barrel of flour from Pittsburgh to Philadelphia amounted to one dollar.[86] Thus coal was hauled in off-seasons when trade at Pittsburgh was dull. This was usually in the summertime when the long, rainless weeks of the Midwest had reduced the flow of the Ohio.

Eastern demand, particularly from the gasworks of Philadelphia and New York, rapidly increased the volume of coal tonnage on the Pennsylvania. The railroad gradually came to look upon coal freight as one of its most profitable sources of revenue. By 1858 the line hauled three hundred thousand tons of coal. Two hundred thousand tons were shipped east and, interestingly enough, the remaining one hundred thousand tons found ready market at Pittsburgh.[87]

The story of the Pennsylvania coal trade would not be complete without reference to the semibituminous coal commerce. Before the beginning of the Civil War, semibituminous, an excellent steam coal, was mined in two separate regions of the Commonwealth. The Blossburg and Barclay coal came from north central Pennsylvania. Broad Top coal was mined in the isolated fields of the south central part

---

[84] *Merchants' Magazine*, XXXIV, 633-34.

[85] *Pennsylvania Senate Journal*, 1838, Appendix to II, 25.

[86] CSHS, Palmer Papers, Coal Notes, October 2, 1856.

[87] *Twelfth Annual Report, Pennsylvania Railroad* (1859), p. 14.

of the State. The Blossburg mines sent their first shipments to market in 1840. By 1861 the mines furnished the western New York area with over a hundred thousand tons a year. Saltworks and other industries were supplied by canal and rail connections between Pennsylvania and New York. In 1856 the Barclay Coal Company began operations nearby in Bradford County. The mines were some thirty-six miles from Waverly, New York. Like the Blossburg coal, most of the Barclay coal was shipped north to New York State. Production, however, was small until new companies entered the region after 1865.[88]

To the south, Broad Top mines were opened in 1855-56 in the midst of a boom reminiscent of the earlier days of anthracite strikes in Schuylkill County. Capital was invested quickly in the excellent steam coal. Mining corporations were formed and towns were erected in what had been wilderness. Soon, a railroad, the Huntingdon and Broad Top, connected the coal fields with the Pennsylvania Railroad and eastern markets. The growth of the area was little short of phenomenal. By 1861 over a quarter of a million tons a year moved from Broad Top to the eastern seaboard.[89]

The Pennsylvania bituminous and anthracite coal trade furnished the young nation with eighty per cent of its mineral fuel. To many contemporaries the energy and comfort which were produced by the black mineral wealth of Pennsylvania ushered in a new age—the "Coal Age." Cephas Grier Childs, editor of the *Philadelphia Commercial List,* once wrote of Pennsylvania: "Here a bountiful Providence has lavished in great profusion his richest gifts. . . . Our state is an *Empire* within itself."[90] Nature, indeed, had been kind to the Commonwealth. Fertile soil and excellent climate had placed the region in the forefront of agriculture. Its rivers provided natural avenues of commerce. Iron ore and limestone contributed to the beginnings of heavy industry. But above all, the early years of America's extraordinary progress toward industrialization and urban growth were made possible by the utilization of the seemingly endless supplies of energy found in the anthracite and bituminous deposits of Pennsylvania's Coal Age Empire.

---

[88] Saward, *The Coal Trade,* 1878, pp. 11-12.

[89] *Bedford Inquirer and Chronicle,* November 23, 1855; Saward, *The Coal Trade,* 1878, p. 14.

[90] HSP, C. G. Childs, Twenty-four Notebooks.

Appendix

TABLE A

Average Retail Prices of Anthracite Coal on the New York Market
per ton of 2,000 lbs.*

Year	Low	High	Year	Low	High
1826	$11.00	$12.00	1843	$ 4.50	$ 6.00
1827	$10.50	$12.50	1844	$ 4.50	$ 6.00
1828	$10.00	$12.00	1845	$ 4.50	$ 6.00
1829	$10.00	$12.00	1846	$ 5.00	$ 7.00
1830	$ 7.00	$12.00	1847	$ 5.00	$ 7.00
1831	$ 6.00	$ 9.00	1848	$ 4.50	$ 6.00
1832	$ 7.50	$16.00	1849	$ 5.00	$ 6.00
1833	$ 5.50	$10.00	1850	$ 5.00	$ 7.00
1834	$ 5.50	$ 6.50	1851	$ 4.25	$ 7.00
1835	$ 5.50	$ 9.00	1852	$ 5.00	$ 7.00
1836	$ 7.00	$11.00	1853	$ 5.00	$ 7.00
1837	$ 8.50	$11.00	1854	$ 6.00	$ 7.50
1838	$ 7.00	$ 9.50	1855	$ 5.50	$ 7.50
1839	$ 6.50	$ 9.00	1856	$ 5.50	$ 6.50
1840	$ 6.00	$ 8.50	1857	$ 6.00	$ 7.00
1841	$ 6.50	$ 9.00	1858	$ 5.00	$ 6.00
1842	$ 5.00	$ 9.00	1859	$ 5.50	$ 5.50
			1860	$ 5.50	$ 6.00

* Compiled from *House Executive Document,* 38th Congress, 1st Session, VI,
1863-64, pp. 362-401.

Note: Prices varied with the season and were not constant. Sometimes the highest
price was in June or July and not at the peak of the winter season.

TABLE B

Tons of Bituminous Coal Consumed, Philadelphia Gas Works*

A. Estimated Amount in Tons		A. Estimated Amount in Tons	
1840	5,000	1848	12,182
1845	6,500	1852	23,000
1846	8,000		

* Compiled from the *Annual Report, PGW,* 1840-1865.

B. Recorded Amount in Tons		B. Recorded Amount in Tons	
1853	25,821	1858	49,040
1854	36,500	1860	64,735
1855	38,158	1862	75,563
1856	44,468	1864	83,744
1857	47,636	1865	90,000

TABLE C

Year	Tons of Anthracite Pig Iron (actual) Pensylvania*	Tons of Anthracite Coal (estimated @ 2 1/5 tons for each ton of pig iron) **
1849	115,000	253,000
1854	208,603	458,926
1855	255,326	562,717
1856	306,972	674,338

* *Twenty-Sixth Annual Report of the Philadelphia Board of Trade,* 1859, pp. 123-124 (figures supplied by J. P. Lesley, Secretary of the American Iron Association).

** Estimate of 2 1/5 is that of C. G. Childs, Editor of *The Philadelphia Commercial List,* found in his Twenty-four Notebooks, HSP.

TABLE D*

Tons of Pennsylvania Pig Iron—1856		All Others U.S.	Total
Anthracite Pig	306,972	87,537	394,509
Charcoal Pig	96,154	252,700	340,854
Coke Pig	39,953	4,528	44,481
Bituminous Pig	8,417	16,656	24,073
Total tons	451,496	361,421	803,917

* *Ibid.,* p. 113.

TABLE E

Approximate Amounts and Expenditures on Wood and Coal Fuel,
Philadelphia and Reading Railroad*

Year	Cords of Wood	Wood Costs Including Cutting & Cording	Tons of Anthracite	Fuel Costs: Locomotives	Fuel Costs: Stationary Engines
1843	15,555				
1844	24,148	$53,396		$3,606	
1845	43,218				
1846	68,006				
1847	90,716	$225,622			$9,764
1848	66,686	157,998		$4,389	8,888
1849	50,264	156,000			9,237
1850	53,997	184,337	3,306		
1851	61,222	224,611	5,640		
1852	52,990	192,256	13,853	$41,118	
1853	38,376	153,570	24,791	52,000	
1854	31,544	143,970	40,178	122,142	
1855	38,140	114,894	52,145	121,498	
1856	20,334	77,940		95,788	
1857					
1858					
1859	15,000		41,000		

* From the *Annual Report.*

TABLE F

Pennsylvania Railroads—Coal and Wood Consumption—1859*

Name	Length of Road in miles	Coal (tons)	Wood (cords)
Allegheny Valley R.R. Co.	45	unknown	unknown
Barclay R.R. & Coal Co.	16	"	"
Beaver Meadow R.R. Co.	20.47	7,611	939
Chestnut Hill R.R.	4.12		1,000
Chartiers Valley R.R.		unfinished	
Catawissa, Williamsport & Erie	65		10,315
Chester Valley R.R. Co.	21.5	unknown	unknown
Cumberland Valley R.R. Co.	52	324	1,612
Delaware, Lackawanna & Western	113	10,200	34,000
Donaldson Improvement R.R. Co.		"not significant"	
East Pa. R.R. Co.	36	not in operation	
Erie & Northeast R.R.	18.5		6,000
Franklin R.R. Co.	22	unfinished	
Gettysburg R.R. Co.	17		350
Hanover Branch R.R. Co.	12.2		1,894
Hempfield R.R. Co.	32	unknown	unknown
Lackawanna and Bloomsburg	80	100	2,600
Lehigh Valley	46	8,000	2,000
Littlestown R.R. Co.	7.25	unknown	
Lehigh & Susquehanna		"not answered"	
Lorbury Creek	5	"not answered"	
Lykens Valley R.R. & Coal Co.	15.5	124	1,450
Mauch Chunk & Sumont Hill	no report—private road of Lehigh Coal & Navigation Co.		
M'Cauly Mt. R.R. Co.	no report—private road of Columbia Coal & Iron Co.		
Mt. Carbon & Port Carbon	2.5	unknown	unknown
Northwestern R.R. Co.	90	in construction	
Philadelphia and Trenton	28.2	no report	
Pennsylvania R.R. Coal Co.	47	unknown	unknown
New York and Erie		"none"	87,000
Pa. R.R. Co.		73,000	59,000
Phila. & Reading R.R. Co.		41,000	15,000
Total given for 1859		179,274	208,192

* Information taken from *Pennsylvania Legislative Documents,* 1860, pp. 549-707 (in pursuance with the Act of April 4, 1859, to make reports to the Auditor General).

Note: See *Pennsylvania Legislative Documents,* 1861, pp. 455-643, for all railroads in Pennsylvania having State charters. Figures for coal and wood, 1860, do not change in ratio to 1859 figures.

TABLE G

Tons of Anthracite Coal Transported by Railroad*

Year	Philadelphia and Reading	Lehigh Valley	Delaware, Lackawanna and Western
1855	2,213,292	8,482	187,000
1856	2,088,903	165,740	305,530
1857	1,709,692	418,235	490,023
1858	1,542,646	471,029	683,411
1859	1,632,932	577,641	829,435
1860	1,946,195	730,641	1,080,227

* F. E. Saward, *The Coal Trade*, 1878, pp. 3-5; *Philadelphia and Reading Railroad, Annual Reports*.

TABLE H

Boston Anthracite Tonnage—Chiefly from the Port of Philadelphia*

1834	76,180	1847	258,093
1835	75,722	1848	274,902
1836	67,186	1849	262,632
1837	80,557	**1850	350,000
1838	71,364	**1851	355,000
1839	90,485	**1852	430,000
1840	73,847	1853	364,888
1841	110,432	1854	369,772
1842	90,276	1855	387,092
1843	117,451	1856	406,963
1844	139,566	1857	370,960
1845	165,422	1858	419,764
1846	168,001		

* R. C. Taylor, *Statistics of Coal*, p. 458; Hazard, *U.S. Register*, VI, January 19, 1842, p. 41; *Boston Daily Advertiser*. Reprinted in the *Twenty-sixth Annual Report of the Philadelphia Board of Trade*, 1859, p. 151.

** Figures do not agree. These are estimated amounts.

TABLE I

Tons of Anthracite Coal at Baltimore*

1851	200,000	1856	266,661
1852	125,000	1857	243,482
1853	183,000	1858	256,105
1854	238,740	1859	268,000 (estimated)
1855	265,921		

* *Miners' Journal*, January 15, 1859, January 14, 1860; *Report of the Philadelphia Board of Trade*, 1859 and 1860.

TABLE J

Coal Received at Cincinnati (Estimated Amounts) *

1853-54	8 million bushels
1854-55	10¼ million bushels
1855-56	7½ million bushels
1857-58	14½ million bushels
1858-59	15 million bushels
1859-60	12⅓ million bushels
1860-61	17½ million bushels

* *Annual Statement of the Trade and Commerce of Cincinnati*, 1858-1861.

TABLE K*

Bituminous Coal at Pittsburgh

1828	1,000,000 bu.—estimated
1833	6,165,480 bu.—returned
1837	11,304,000 bu.—estimated
1842	12,760,000 bu.—estimated
1846	19,000,000 bu.—announced
1856	53,782,159 bu.—returned

Pittsburgh Exports—1856

Ohio River	16,300,159 bu.
Pa. Railroad	3,372,000 bu.
Pa. Canals	560,000 bu.
Cleveland & Pittsburgh R.R.	140,000 bu.

* *Pittsburgh Quarterly Trade Circular*, I, 1859, pp. 29-30.

TABLE L

Pennsylvania Bituminous Coal at New Orleans**

Year	Barrels*	Net Tons	Year	Barrels*	Net Tons
1816		60	1846	153,740	15,374
1830	4,200	420	1847	229,900	22,990
1833	21,920	2,192	1848	194,000	19,400
1835	47,343	4,734	1849	232,500	23,250
1836	71,425	7,143	1850	296,500	29,650
1837	51,330	5,133	1851	40,000	4,000
1838	94,400	9,440	1852	829,100	82,910
1839	85,762	8,576	1853	456,600	45,660
1840	61,406	6,141	1854	545,500	54,550
1841	87,641	8,764	1855	1,118,000	111,800
1842	47,564	4,756	1856	192,000	19,200
1843	73,850	7,385	1857	619,700	61,970
1844	45,203	4,520	1858	677,000	67,700
1845	140,200	14,020	1859	965,000	96,500
			1860	1,685,000	168,500
			1861	1,510,000	151,000

* 1 barrel—200 lbs. or 2½ bushels of coal. 10 barrels—2000 lbs. or 1 net ton.

** Eavenson, *American Coal Industry*, pp. 389-90.

TABLE M

Bituminous Coal Received at Louisville*

1841	3,200 net tons
1843	24,520 net tons
1852	101,862 net tons
1860	240,000 net tons

* Eavenson, *American Coal Industry*, p. 404.

TABLE N

Monongahela Navigation Company

	Bushels*	Tons**
1845	4,605,185	184,200
1846	7,778,911	311,156
1847	9,645,127	385,805
1848	9,819,361	392,774
1849	9,708,507	398,340
1850	12,297,967	491,918
1851	12,521,228	490,850

Combined Coal Shipments Down Monongahela and
Youghiogheny River Improvements

1852	14,630,841	585,233
1853	15,716,367	628,654
1854	17,331,946	693,278
1855	22,234,009	889,360
1856	8,584,095	353,364
1857	28,973,596	1,158,939
1858	25,696,669	1,027,866
1859	28,286,671	1,131,467
1860	37,947,732	1,517,909

* Wall, *Second Geological Survey of Pennsylvania, Report of Progress, K4*.

** Saward, *The Coal Trade*, 1878, p. 20. Tonnage estimated as 25 bushels to the ton of 2000 pounds.

Bibliography

MANUSCRIPT SOURCES

Historical Societies

Colorado State Historical Society
  The William Jackson Palmer Papers

Chester County Historical Society
  Pennypacker Papers

Historical Society of Schuylkill County
  Baird Halberstadt Papers
  Eyre-Ashurst Papers
  Parry Papers

Historical Society of Pennsylvania
  Henry C. Carey Papers, Edward Carey Gardiner Collection
  Buck, W. J., Early Discovery of Coal, MS
  Childs, C. G., Twenty-four Notebooks, 1845-1865
  Fell, Jesse, to Johnathan Fell, Wilkes-Barre, December 1, 1826
  James, T. C., A Reminiscence, MS
  Malin, Joshua, to Gerard Ralston, Esq., April 20, 1827
  White, Josiah, to Governor Joseph Ritner, July 11, 1837

Private Collections

Baltimore and Ohio Railroad Company, Minute Books, 1836-1842
Delaware and Hudson Minute Books, 1825-1865
Donaghy and Sons, Philadelphia, Coal Yard Receipts, 1837-1866
F. and S. Jones, Germantown, Ledgers, 1855-1865

PRINTED DOCUMENTS

Senate Document, No. 386, 28th Congress, 1st Session.
Senate Executive Document, No. 74, 32nd Congress, 1st Session.
Senate Executive Document, No. 31, 32nd Congress, 2nd Session.
House Executive Document, No. 65, 32nd Congress, 2nd Session, *Report of the Commissioner of Patents,* 1852
House Executive Document, No. 96, 36th Congress, 1st Session, 1860.
House Executive Document, 38th Congress, 1st Session, VI, 1863-1864.
Pennsylvania Archives, 4th Series, Papers of the Governors, I-XII.
Pennsylvania Senate Journal, 1803-1861.
Pennsylvania House Journal, 1803-1861.
Pennsylvania Executive Documents, 1845-1846.
Pennsylvania Legislative Documents, 1853-1861.
Documents Relating to the Manufacture of Iron in Pennsylvania. Published on behalf of the Convention of Iron Masters, which met in Philadelphia, December 20, 1849. Philadelphia, 1850.

169

Manufactures of the United States in 1860; Compiled from the Original Returns of the Eighth Census. Washington, 1865.

Official Records of the Union and Confederate Navies in the War of the Rebellion. Washington, 1894.

OFFICIAL REPORTS

Blizard, John, James Neil, and F. C. Houghton. *Value of Coke, Anthracite and Bituminous Coal for Generating Steam in a Low-Pressure Cast Iron Boiler.* (Technical Paper 303, Department of the Interior, Bureau of Mines.) Washington, 1922.

Cohen, Mendes. *Report on Coke and Coal Used for Passenger Trains on the Baltimore and Ohio Railroad.* 1854.

Cresson, J. C. *Report to the Trustees of the Philadelphia Gas Works,* August 8, 1845.

Johnson, W. R. *Report on an Examination of the Mines, Iron Works, and Other Property Belonging to The Clearfield Coke and Iron Company.* Philadelphia, 1839.

Merrick, S. V. *Report on the Gas Works of Europe,* December, 1834.

Neal, D. A. *Reports Made to the Managers of the Philadelphia and Reading Railroad Company.* 1849-1850.

Roberts, W. M. *Report on the Erie Extension to the Board of Canal Commissioners, Pennsylvania House Journal,* Appendix to II. 1840.

Wall, J. S. *Report on the Coal Mines of the Monongahela River Region, From the West-Virginia State Line to Pittsburgh, Including the Mines on the Lower Youghioheny River. Second Geological Survey of Pennsylvania, Report of Progress K4.* Harrisburg, 1884.

Weeks, J. D. *Report on the Manufacture of Coke, Tenth Census,* X. 1880.

Wetherill, C. M. *Report on the Iron and Coal of Pennsylvania.* 1850.

Whistler, G. W., Jr. *Report Upon the Use of Anthracite Coal in Locomotive Engines on the Reading Railroad, Made to the President of the Philadelphia and Reading Railroad Company, April 20, 1849.* Baltimore, 1849.

Wilkes, Charles. *Report on the Examination of the Deep River District, North Carolina.* 1858.

Report to the Select and Common Councils of the City of Philadelphia by the Committee on Lighting the City with Gas. 1833.

Report of the Committee of the Senate of Pennsylvaina upon the Subject of the Coal Trade. Harrisburg, 1834.

Report of the Engineers of the Philadelphia and Reading Railroad Company with Accompanying Documents, etc. Philadelphia, 1838.

Report to the Legislature of Pennsylvania Containing a Description of the Swatara Mining District. Harrisburg, 1839.

Report of the Board of Managers of the Lehigh Coal and Navigation Company, January 14, 1839.

Report of a Committee to the Iron and Coal Association of the State of Pennsylvania. Philadelphia, 1846.

Annual Report on the Geological Exploration of the State of Pennsylvania, 1836-1839.

Annual Report of the Trustees of the Philadelphia Gas Works, 1836-1865.

Annual Report of the Philadelphia Board of Trade, 1850-1860.

Annual Report of the Schuylkill Navigation Company, 1855.

Annual Statement of the Trade and Commerce of Cincinnati, 1858-1860.

Annual Report of the Chamber of Commerce of the State of New York, 1858-1860.

7th Annual Report of the American Institute of the City of New York, 1849.

Annual Report of the Board of Managers of the Delaware and Hudson Canal Company to the Stockholders, 1825-1865.
Annual Report of the Baltimore and Ohio Railroad Company, 1827-1865.
Annual Report of the Philadelphia and Reading Railroad Company, 1838-1865.
Annual Report of the Monongahela Navigation Company, 1840-1850. Reprinted in the *Pennsylvania House* and *Senate Journal.*
Annual Report of the Pennsylvania Railroad Company, 1847-1861.
Annual Report of the President and Directors to the Stockholders of the Cleveland and Pittsburgh Railroad, 1849-1856.
Annual Report of the Board of Directors of the North Pennsylvania Railroad Company, 1854.
Annual Report of the Cincinnati and Pittsburgh Railroad, 1856-1860.
Annual Report of the Pittsburgh, Fort Wayne and Chicago Railroad, 1858-1859.

PUBLISHED DIARIES, MEMOIRS, LETTERS, AND COLLECTED WORKS

Beckerath, E. V., Karl Goeser, *et al.* (eds.). *Friedrich List, Werke,* II. Berlin, 1928-1933.
Bremer, Fredrika. *The Homes of the New World; Impressions of America,* I. New York, 1854.
Carey, Henry C. *Works of Henry C. Carey,* XXIX. Philadelphia, 1867.
Clothier, I. H. (comp.). *Letters, 1853-1868, General William Jackson Palmer.* Philadelphia, 1906.
Elliott, R. S. *Notes Taken in Sixty Years.* St. Louis, 1883.
Emmerson, J. C., Jr. (ed.). *Steam Navigation in Virginia and Northeastern North Carolina Waters, 1826-1836.* Portsmouth, Virginia, 1949.
Morse, J. T., Jr. (ed.). *The Diary of Gideon Welles,* III. Boston, 1910.
Nevins, Allen (ed.). *The Diary of Philip Hone,* I, II. New York, 1927.
Pope-Hennessy, Una (ed.). *The Aristocratic Journey.* New York, 1931.
Thwaites, R. G. (ed.). *Early Western Travels 1748-1846,* III, IV, VIII, XIII. Cleveland, 1904-06.

NEWSPAPERS

Banner of the Constitution, Philadelphia, 1829-1832.
Bedford Inquirer and Chronicle, 1855-1858.
Cist's Weekly Advertiser, Cincinnati, 1845-1851.
Republican Farmer and Democratic Journal, Wilkes-Barre, 1846-1850.
Boston Daily Advertiser, 1829-1831.
Miners' Journal, Pottsville, 1826-1861.
New York Commercial Advertiser, 1836.
New York Journal of Commerce, 1835-1836.
The New York Times, February 14, 1954.
The Pennsylvanian, Philadelphia, 1840, 1842, 1846.
Pittsburgh Gazette, 1854-1860.
Poulson's American Daily Advertiser, Philadelphia, 1831-1832.

PERIODICALS

American Journal of Science and Arts, 1848.
American Railroad Journal, 1832-1860.
De Bow's Review and Industrial Resources, Statistics, etc., 1853.
Journal of the Franklin Institute, 1826-1860, 1874.

Memoirs of the Philadelphia Society for Promoting Agriculture, Containing Communications on Various Subjects in Husbandry and Rural Affairs, III. Philadelphia, 1814.

Niles' Weekly Register, 1814-1816, 1828.

Niles' National Register, 1839-1849.

The North American Review, 1836.

The Pittsburgh Quarterly Trade Circular, 1857.

Scientific American, 1854-1860.

Fisher, Redwood (ed.). *Fisher's National Magazine and Industrial Record,* 1845-1846.

Hazard, Samuel (ed.). *The Register of Pennsylvania,* 1828-1836.

————. *United States Commercial and Statistical Register,* 1839-1842.

Hunt, Freeman (ed.). *The Merchants' Magazine and Commercial Review,* 1839-1860.

Saward, F. E. (ed.). *The Coal Trade,* 1877-1878.

Tenney, W. B. (ed.). *The Mining Magazine,* 1854.

ADDRESSES AND PAMPHLETS

Brown, R. W. "Some Aspects of Early Railroad Transportation in Pennsylvania." Address delivered before the Pennsylvania Historical Association, Dickinson College, Carlisle, Pennsylvania, October 21, 1949.

————. "Friedrich List—the Father of German Railroads: His Residence in Dauphin and Schuylkill Counties, Pennsylvania." Address delivered before the Historical Society of Dauphin County, Harrisburg, Pennsylvania, September 18, 1950.

Bull, Marcus. *An Answer to "A short reply to 'A defence of the experiments to determine the comparative values of the Principal varieties of fuel, etc.' by one of the committee of the American Academy."* Philadelphia, 1828.

Felton, F. E. *Mineral Fuel for Locomotives.* Philadelphia, 1857.

Fox, D. R. *Dr. Eliphalet Nott (1773-1866) and the American Spirit, A Newcomen Address.* Princeton, 1944.

Rhoads, W. W. *When the Railroad Came to Reading! A Newcomen Address.* New York, 1948.

Sanderson, J. M. *A Letter on the Present Condition and Future Prospects of the Reading Railroad in January 1855.* Philadelphia, 1855.

Steele, J. D. *Letter to the President of the Philadelphia and Reading Railroad Company on the Canals and Railroads for Transporting Anthracite Coal.* Philadelphia, 1855.

The Reading Railroad: Its Advantages for the Cheap Transportation of Coal, As Compared with Schuylkill Navigation and Lehigh Canal. Philadelphia, 1839.

Boardman's Coal-Burning Locomotive Boiler Company.

Cannelton, Perry County, Indiana, at the Intersection of the Eastern Margin of the Illinois Coal Basin, by the Ohio River; Its Natural Advantages as a Site for Manufacturing. Published by the American Cannel Coal Company. Louisville, 1850.

Essay on the Manufacture of Iron with Coke. Lewistown, Pennsylvania, 1838.

Facts Illustrative of the Character of The Anthracite or Lehigh Coal, found in the Great Mines of Mauch Chunk, in possession of the Lehigh Coal and Navigation Company, With Certificates from Various Manufacturers, proving its decided superiority over Every Other Kind of Fuel. Philadelphia, 1824.

Memorial of a Convention of Citizens of the Commonwealth for Aid to Enlarge the Union Canal, Read in the Senate, January 17, 1839. Harrisburg, 1839.

Observations On the Rhode Island Coal and Certificates With Regard to its Qualities, Values, and Various Uses. 1814.
Pennsylvania Coal and Carriers. Philadelphia, 1852. Essays reprinted from the Philadelphia *North American*.
Pennsylvania Our Internal Improvements. (n.d.).
Pittsburgh, Her Advantageous Position and Great Resources, as a Manufacturing and Commercial City, Embraced in a Notice of a Sale of Real Estate. Pittsburgh, 1845.
The Voice of Lancaster County Upon the Subject of a National Foundry. Lancaster, 1839.

UNPUBLISHED DISSERTATION

Shegda, Michael. "History of the Lehigh Coal and Navigation Company to 1840." Unpublished dissertation for the Ed. D. at Temple University, 1952.

SECONDARY WORKS

Books

A Century of Reading Motive Power. Philadelphia, 1941.
Albion, R. G. *The Rise of New York Port.* New York, 1939.
Ambler, C. H. *A History of Transportation in the Ohio Valley.* Glendale, California, 1932.
Bezanson, Anne, R. D. Gray and Miriam Hussey. *Wholesale Prices in Philadelphia 1784-1861,* I, II. Philadelphia, 1936.
Bishop, J. L. *A History of American Manufactures from 1608 to 1860,* II. Philadelphia, 1864.
Bogen, J. I. *The Anthracite Railroads.* New York, 1927.
Boucher, J. N. *History of Westmoreland County, Pennsylvania,* I. New York, 1906.
Bowen, Eli. *The Pictorial Sketch-Book of Pennsylvania.* Philadelphia, 1853.
————— (ed.). *The Coal Regions of Pennsylvania.* Pottsville, 1848.
Brown, G. T. *The Gas Light Company of Baltimore, A Study of Natural Monopoly.* Baltimore, 1936.
Brown, W. H. *The History of the First Locomotives in America.* New York, 1871.
Bruce, Kathleen. *Virginia Iron Manufacture in the Slave Era.* New York, 1931.
Cooper, Thomas. *Some Information Concerning Gas Lights.* Philadelphia, 1816.
Coxe, Tench. *A View of the United States of America.* Philadelphia, 1794.
Craig, N. B. *History of Pittsburgh.* Pittsburgh, 1851.
Derby, George, M. D. *Anthracite and Health.* Second edition enlarged, Boston, 1868.
Eavenson, H. N. *The First Century and a Quarter of American Coal Industry.* Pittsburgh, 1942.
Fisher, J. S. *A Builder of the West, the Life of General William Jackson Palmer.* Caldwell, Idaho, 1939.
Freedley, E. T. *Philadelphia and Its Manufactures.* Philadelphia, 1859.
Fulton, John. *Coke.* Scranton, 1905.
Hirst, M. E. *Life of Friedrich List and Selections from His Writings.* New York, 1909.
Hunter, L. C. *Steamboats on the Western Rivers, an Economic and Technological History.* Cambridge, 1949.
Jones, Samuel. *Pittsburgh in the Year 1826, Containing Sketches Topographical, together with a Directory of the City, and a View of its various Manufactures, Population and Improvements.* Pittsburgh, 1826.

Johnson, W. R. *Notes on the Use of Anthracite in the Manufacture of Iron With Some Remarks on Its Evaporative Power.* Boston, 1841.

————. *The Coal Trade of British America with Researches on the Characters and Practical Values of American and Foreign Coal.* Philadelphia, 1850.

Jones, C. L. *The Economic History of the Anthracite-Tidewater Canals.* Philadelphia, 1908.

Klein, F. S. *Lancaster County, 1841-1941.* Lancaster, 1941.

Lesley, J. P. *Manual of Coal and Its Topography.* Philadelphia, 1856.

————. *The Iron Manufacturers Guide to the Furnaces, Forges and Rolling Mills of the United States, etc.* New York, 1859.

Livingood, J. L. *The Philadelphia-Baltimore Trade Rivalry 1780-1860.* Harrisburg, 1947.

Lunge, George. *Coal-Tar and Ammonia,* I, III. New York, 1916.

Matthews, William. *An Historical Sketch of the Original Progress of Gas-Lighting.* London, 1832.

Mease, James. *The Picture of Philadelphia.* Philadelphia, 1811.

Morton, Eleanor. *Josiah White, Prince of Pioneers.* New York, 1946.

Nicolls, W. J. *The Story of American Coals.* Philadelphia, 1897.

Peirce, J. H. *Fire on the Hearth, the Evolution and Romance of the Heating Stove.* Springfield, 1951.

Reiser, C. E. *Pittsburgh's Commercial Development 1800-1850.* Harrisburg, 1951.

Renwick, James. *Treatise on the Steam Engine.* New York, 1830.

Scharf, J. T., and Thompson Westcott. *History of Philadelphia,* I, III. Philadelphia, 1884.

Sprout, Harold and Margaret. *The Rise of American Naval Power.* Princeton, 1946.

Swank, J. W. *Introduction to the History of Iron Making and Coal Mining in Pennsylvania.* Philadelphia, 1878.

————. *History of the Manufacture of Iron in all Ages.* Philadelphia, 1892.

Taylor, R. C. *Statistics of Coal.* Philadelphia, 1855.

Tyler, D. B. *Steam Conquers the Atlantic.* New York, 1939.

Van Doren, Carl. *Benjamin Franklin.* New York, 1939.

ARTICLES

Bining, A. C. "The Rise of Iron Manufacture in Western Pennsylvania," *The Western Pennsylvania Historical Magazine,* XVI (November, 1933), 235-256.

————. "Ironmen in Quest of Fuel," *Steelways,* X (August, 1954).

Brunot, F. R. "Improvements of the Ohio River," *Journal of the Franklin Institute,* LXVII (1874).

Bull, Marcus. "Experiments to determine the comparative quantities of Heat evolved, in the combustion of the principal varieties of Wood and Coal used in the United States, for fuel; and also, to determine the comparative quantities of Heat lost by the ordinary apparatus made use of for their combustion," Read April 7, 1826, *Transactions of the American Philosophical Society, Held at Philadelphia for Promoting Useful Knowledge,* III, new series (1830).

Crane, George. "On the Smelting of Iron with Anthracite Coal," *Journal of the Franklin Institute,* XXV (1838).

Eavenson, H. N. "The Early History of the Pittsburgh Coal Bed," *The Western Pennsylvania Historical Magazine,* XXII (September, 1939), 165-176.

Griffith, William. "The Proof That Pennsylvania Anthracite Coal Was First Shipped from the Wyoming Valley," *Proceedings and Collections of the Wyoming Historical and Geological Society,* XIII (1913-14), 65-71.

Haldeman, H. L. "The First Furnace Using Coal," *Lancaster County Historical Society Papers,* I (1896).

Hunter, L. C. "Influence of the Market upon Technique in the Iron Industry in Western Pennsylvania up to 1860," *Journal of Economic and Business History,* I (February, 1929).

——————. "Seasonal Aspects of Industry and Commerce Before the Age of Big Business." *Smith College Studies in History,* XIX (1933-34).

Murdoch, William. "An Account of the Application of Gas from Coal to Economical Purposes," *Abstracts of the Papers Printed in the Philosophical Transactions of the Royal Society of London From 1800 to 1830,* I (1800-14). London, 1832.

Neilson, J. B. "On the Hot Air Blast," *Transactions of the Institution of Civil Engineers,* I (1836).

Peterman, C. E. "Early House Warming by Coal Fires," *Journal of the Society of Architectural Historians,* IX (December, 1950).

Rogers, H. D. and Bache. "Analysis of some of the Coals of Pennsylvania," *Journal of the Academy of Natural Sciences,* VIII (1834).

Thurston, G. H. "Pittsburgh As It Is," *The Pittsburgh Quarterly Trade Circular,* I (October, 1857).

Woodhouse, James. "Experiments and Observations on the Lehigh Coal," *The Philadelphia Medical Museum,* I (1805).

Index